Design and Control of Power Converters 2020

Design and Control of Power Converters 2020

Editor

Manuel Arias

MDPI • Basel • Beijing • Wuhan • Barcelona • Belgrade • Manchester • Tokyo • Cluj • Tianjin

Editor
Manuel Arias
University of Oviedo
Spain

Editorial Office
MDPI
St. Alban-Anlage 66
4052 Basel, Switzerland

This is a reprint of articles from the Special Issue published online in the open access journal *Energies* (ISSN 1996-1073) (available at: https://www.mdpi.com/journal/energies/special_issues/DC_PC_2020).

For citation purposes, cite each article independently as indicated on the article page online and as indicated below:

LastName, A.A.; LastName, B.B.; LastName, C.C. Article Title. *Journal Name* **Year**, *Volume Number*, Page Range.

ISBN 978-3-0365-0702-6 (Hbk)
ISBN 978-3-0365-0703-3 (PDF)

Cover image courtesy of pixabay.com user Blickpixel.

© 2021 by the authors. Articles in this book are Open Access and distributed under the Creative Commons Attribution (CC BY) license, which allows users to download, copy and build upon published articles, as long as the author and publisher are properly credited, which ensures maximum dissemination and a wider impact of our publications.
The book as a whole is distributed by MDPI under the terms and conditions of the Creative Commons license CC BY-NC-ND.

Contents

About the Editor vii

Preface to "Design and Control of Power Converters 2020" ix

Claudio Adragna, Giovanni Gritti, Santi Agatino Rizzo and Giovanni Susinni
Distortion Due to the Zero Current Detection Circuit in High Power Factor Quasi-Resonant Flybacks
Reprinted from: *Energies* **2021**, *14*, 395, doi:10.3390/en14020395 1

Paweł Górecki and Krzysztof Górecki
Analysis of the Usefulness Range of the Averaged Electrothermal Model of a Diode–Transistor Switch to Compute the Characteristics of the Boost Converter
Reprinted from: *Energies* **2021**, *14*, 154, doi:10.3390/en14010154 27

Ming Liu, Zetao Li and Xiaoliu Yang
A Universal Mathematical Model of Modular Multilevel Converter with Half-Bridge
Reprinted from: *Energies* **2020**, *13*, 4464, doi:10.3390/en13174464 43

Ahmed H. Okilly, Namhun Kim and Jeihoon Baek
Inrush Current Control of High Power Density DC–DC Converter
Reprinted from: *Energies* **2020**, *13*, 4301, doi:10.3390/en13174301 61

Goh Teck Chiang and Takahide Sugiyama
Methods of Modulation for Current-Source Single-Phase Isolated Matrix Converter in a Grid-Connected Battery Application
Reprinted from: *Energies* **2020**, *13*, 3845, doi:10.3390/en13153845 83

Sarah Saeed, Ramy Georgious and Jorge Garcia
Modeling of Magnetic Elements Including Losses—Application to Variable Inductor [†]
Reprinted from: *Energies* **2020**, *13*, 1865, doi:10.3390/en13081865 105

G. Kiran Kumar, E. Parimalasundar, D. Elangovan, P. Sanjeevikumar, Francesco Iannuzzo and Jens Bo Holm-Nielsen
Fault Investigation in Cascaded H-Bridge Multilevel Inverter through Fast Fourier Transform and Artificial Neural Network Approach
Reprinted from: *Energies* **2020**, *13*, 1299, doi:10.3390/en13061299 125

Mariam Saeed, María R. Rogina, Alberto Rodríguez, Manuel Arias and Fernando Briz
SiC-Based High Efficiency High Isolation Dual Active Bridge Converter for a Power Electronic Transformer
Reprinted from: *Energies* **2020**, *13*, 1198, doi:10.3390/en13051198 145

Maria R. Rogina, Alberto Rodrigusez, Diego G. Lamar, Jaume Roig, German Gomez and Piet Vanmeerbeek
Analysis of Intrinsic Switching Losses in Superjunction MOSFETs Under Zero Voltage Switching
Reprinted from: *Energies* **2020**, *13*, 1124, doi:10.3390/en13051124 163

About the Editor

Manuel Arias received an M.Sc. in electrical engineering from the University of Oviedo, Spain, in 2005, and a Ph.D. from the same university in 2010. In 2007, he joined the University of Oviedo as an Assistant Professor and, since 2016, he has been an Associate Professor at the same university. His research interests include AC–DC and DC–DC converters, battery-cell equalizers, LED lighting, and aerospace applications.

Preface to "Design and Control of Power Converters 2020"

In terms of research, power electronics is one of the most prolific fields in the world of electronics. One of the main reasons for this is its relevance in present-day society, which is increasingly concerned with energy savings and greener energy production. This scenario constitutes a powerful catalyst for research, not only increasing the amount of interesting ideas, solutions, and studies, but also the number of topics that have emerged under the umbrella of power electronics. This can be observed in well-established research topics as varied as renewable energies, battery management, and electric traction coexisting, or even merging, with more recent topics, such as LED lighting or micro- and nano-grids. These topics can be considered as established compared to others like wide band-gap devices and electric vehicles, where research is still incipient. In all of the aforementioned topics, in addition to others, the design and control of power converters plays a key role. In this book, representative papers that focus on well-established topics, as well as more recent ones, can be found. This mixture will foster new ideas for readers and help researchers detect solutions that can be migrated from one topic to another, making this book a relevant milestone for any power electronics engineer.

Manuel Arias
Editor

Article

Distortion Due to the Zero Current Detection Circuit in High Power Factor Quasi-Resonant Flybacks

Claudio Adragna [1], Giovanni Gritti [1], Santi Agatino Rizzo [2,*] and Giovanni Susinni [2]

[1] Industrial and Power Conversion Division Application Laboratory, STMicroelectronics S.R.L., 20864 Agrate Brianza, Italy; claudio.adragna@st.com (C.A.); giovanni.gritti@st.com (G.G.)
[2] Department of Electrical, Electronic and Computer Engineering, University of Catania, 95125 Catania, Italy; giovanni.susinni@unict.it
* Correspondence: santi.rizzo@unict.it; Tel.: +39-09-5738-2308

Abstract: In a high-power factor quasi-resonant Flyback, an ideal zero current detection (ZCD) circuit and control circuitry enable the power switch turn-on in the exact instant a zero ringing current is reached after demagnetization. A nonzero current at the turn-on instant affects the input current shape and; consequently, affects its Total Harmonic Distortion (THD). This paper firstly deeply analyzes the effect on the distortion due to a nonideal ZCD circuit. After, some typical implementations of the ZCD circuit and their effect on the THD are analyzed, identifying their pros and cons. Finally, some experimental results are obtained to validate the analytical investigation.

Keywords: converter control; power factor correction; total harmonic distortion; flyback; solid-state lighting

1. Introduction

The Flyback topology circuit represents one of the most attractive dc-dc converters used for low and medium power applications. To date, the Flyback converters have been widely employed as USB chargers for cell phones, notebooks, LCD TVs, and LED drivers [1–6]. The key factors that make this solution very popular are related to the simple design of the conversion stage, high efficiency, inexpensive cost of the components, possible multiple isolated output stages, and high output voltage.

It is worth remembering that the Flyback topology has been widely studied in the literature, where several advantageous results with respect to other converter topologies have been pointed out. In this perspective, the authors in [7] have compared the performance between a flyback, a buck-boost, and a hybrid solution [8–14] in terms of some key factors (e.g., cost, efficiency, step-down ability, etc.). The comparison has highlighted the superior performance of the flyback converter, which makes it the preferred choice for countless dc-dc power applications.

The flyback converter can be summarized into two different operating modes. When the current in the secondary windings goes to zero before the OFF time of the power switch, the Flyback operates in discontinuous conduction mode (DCM). On the other hand, when the current in the primary windings is greater than zero before the OFF time is complete, the converter operates in the continuous conduction mode (CCM) [15].

In many applications, the flyback converter can be used with a feedback control loop, with the aim to sense the voltage and current variations of the output. Generally, an optocoupler is used to sense the output voltage and in addition, it provides electrical isolation (SSR—secondary side regulation). Nevertheless, the use of an optocoupler presents a few disadvantages [16]. Indeed, the current transfer ratio may be subjected to variation due to the temperature, and furthermore, the optocoupler introduces an undesired pole in the feedback loop, which may lead to an instability of the converter.

On the other hand, the adoption of the constant-current primary sensing regulation (CC-PSR) technique [3,4], the feedback loop at the output(secondary) side is not used and consequentially, the output voltage can be adjusted by using only the control method of the electrical quantities in the primary side. This leads to a save cost and the power losses are strongly reduced. More specifically, in this configuration, only the auxiliary winding is used to sense the output voltage. It is worth underlining that this approach brings greater safety and reliability. As a disadvantage, the primary-side control may result in less accuracy in comparison to the feedback conventional method.

The high-power-factor (Hi-PF) flyback converter is one of the most popular topologies used in low-medium power applications (e.g., LED drivers that feed from the ac power line [17–20]). It has been widely used due to its low input current distortion (low THD) and it guarantees high safety isolation. Furthermore, it is easily suitable for CC-PSR operation [21–23]. The Hi-PF flyback converters can be used with a fixed switching frequency and in the DCM mode (FF-DCM) [24,25]. This mode operation is theoretically able to obtain unity power factor and zero THD.

Furthermore, the Hi-PF flyback converters can also be used in the Quasi-Resonant (QR) mode, where the switching period depends on transformer demagnetization. It presents several advantages compared to FF-DCM such as valley-switching, (almost zero-voltage switching, ZVS) and low EMI emissions. However, its standard implementation guarantees a sinusoidal envelope of the peaks of the primary current. This cannot achieve a very low THD of the input current, as widely discussed in [22–27]. However, more recently, an enhanced QR control method [28] able to provide Hi-PF QR flyback converters with the ability to ideally get a sinusoidal input current has been disclosed.

In this paper, the contribution to the distortion of the input current in a Hi-PF Flyback converter due to a nonzero current at the turn-on instant of the power switch in a switching cycle has been discussed in depth.

Firstly, the general trend of the impact of a nonzero current at the turn-on instant on the input current shape and, ultimately, on its THD is analyzed with both the traditional (QR) method and the enhanced QR (EQR) method. Secondly, this impact is analyzed more in detail with reference to some typical implementations of the zero current detectors (ZCD) circuit responsible for determining the turn-on instant of the power switch.

It is worth underlining that both the quantitative and the qualitative effects of a nonzero initial current due to the implementation of the zero-current detection circuit have not been yet documented in the literature.

Finally, several experimental results showing the aforementioned contributions in a couple of prototypes of Hi-PF Flyback converter controlled with both the traditional QR method and the enhanced QR method are presented.

2. Review of the Hi-PF QR Flyback Converter and Its Control Methods

A flyback converter (whether Hi-PF or not) is said to be QR-operated when the turn-on of the power switch (usually a MOSFET) is synchronized to the instant when the transformer demagnetizes (i.e., as the secondary current becomes zero), normally after an appropriate delay. In this way, the turn-on can be commanded on the valley of the drain voltage ringing that follows the demagnetization, thus with minimum turn-on losses. For this reason, this operation is often termed "valley-switching".

A Hi-PF QR flyback converter, whose principle schematic and relevant key waveforms are shown in Figures 1 and 2 respectively, is powered from the ac power line with no energy reservoir capacitor after the input bridge rectifier (C_{in} serves as a high-frequency smoothing filter). Thus, its input voltage is essentially a rectified sinusoid ($V_{IN}(\theta) = V_{PK} |sin(\theta)|$) and the current $I_{AC}(\theta)$ drawn from the power line is sinusoidal-like (the rectified input current $I_{IN}(\theta)$ downstream the bridge is $|I_{AC}(\theta)|$. In these expressions and in the following discussion, V_{PK} is the peak line voltage, $\theta = 2p f_L t$ is the instantaneous phase angle of the line voltage, f_L is the line frequency. Note also that uppercase subscripts will refer to quantities considered on a line cycle time scale, lowercase subscripts to quantities considered on a switching cycle time scale. Circuit parameters have a lowercase subscript, dc quantities do not have subscripts.

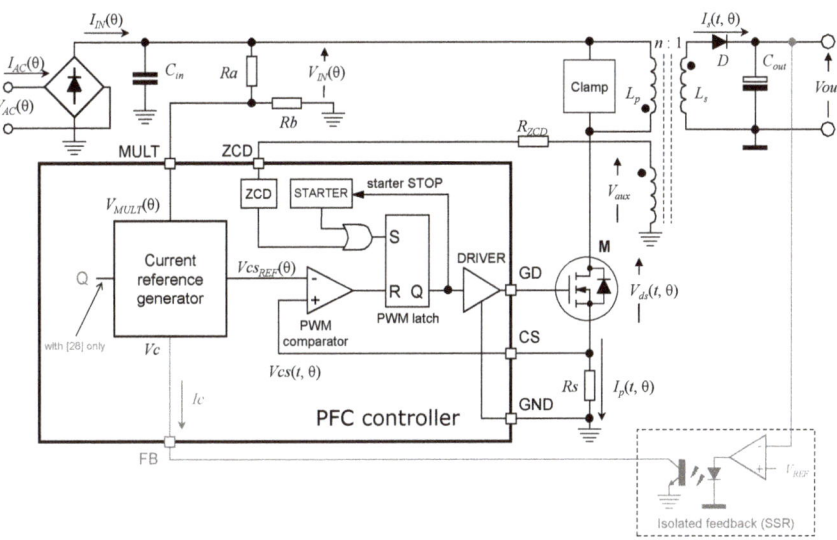

Figure 1. Principle schematic of a Hi-PF QR flyback converter.

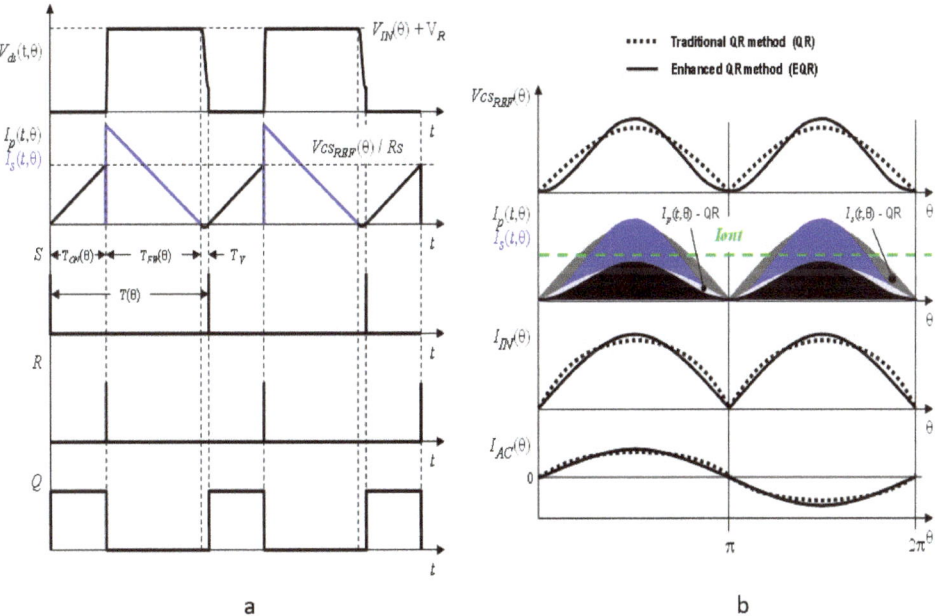

Figure 2. Key waveforms of the circuit in Figure 1; switching cycle time scale (a), line cycle time scale (b). In the second diagram from top on the right-hand side, black and blue shapes are the envelopes of primary and secondary current with EQR control; light grey and dark grey shapes are the same with QR control.

With no loss of generality, whichever type of feedback -SSR or PSR- is used, it is possible to assume that the error signal is processed producing a control voltage V_c that controls the input-to-output power flow. Being a Hi-PF system, the open-loop bandwidth of the overall control loop is very narrow—typically below 20 Hz—and under steady-state operation, Vc can be regarded as a dc level, at least to a first approximation.

Considering peak current mode control, the turn-off of the power switch is determined by the current sense signal reaching the value programmed by the control loop that regulates $Vout$ or $Iout$ via Vc. This value is set by the reference $Vcs_{REF}(\theta)$ output by the "Current reference generator" block that receives at its inputs the control voltage Vc, the voltage $V_{MULT}(\theta)$—a scaled-down image of the input voltage $V_{IN}(\theta)$ that serves as a sinusoidal template—and, in case, the output Q of the PWM latch.

In fact, $Vcs_{REF}(\theta)$ is fed into the inverting input of the PWM comparator that receives the voltage $Vcs(t,\theta)$ on the other input. $Vcs(t,\theta)$ is sensed across the sense resistor R_s, which is proportional to the instantaneous current $I_p(t,\theta)$ flowing through the primary winding L_p and the power switch M when this is in the on state. Assuming that the PWM latch is set (and M turned on) at $t = 0$, the current $I_p(t,\theta)$ will be ramping up linearly and so will do $Vcs(t,\theta)$; at the instant $t = T_{ON}$, when $Vcs(T_{ON}, \theta) = Vcs_{REF}(\theta)$, i.e., $I_p(T_{ON}, \theta) = Vcs_{REF}(\theta)/R_s$, the PWM comparator resets the PWM latch, thus switching off M.

As M is switched off, most of the energy stored in L_p is transferred to the secondary winding L_s so that current starts flowing through L_s and D, dumping this energy into the output capacitor C_{out} and the load. As L_s is completely demagnetized (i.e., the current through L_s zeroes) the diode D opens. The drain voltage V_{ds}, which was fixed at $V_{IN}(\theta)+V_R$ ($V_R = nVout$) while D was conducting, starts oscillating around the instantaneous line voltage $V_{IN}(\theta)$ due to the resonance of the parasitic capacitance of the drain node (C_{ds}) with L_p. The quick drain voltage fall that marks the onset of this ringing is coupled to the ZCD block in the controller through the auxiliary winding L_{aux} and the resistor R_{zcd}. The ZCD block releases a pulse as it detects the negative-going edge and this pulse sets the PWM latch and turns on the power switch M, starting a new switching cycle.

Therefore, the shape of $Vcs_{REF}(\theta)$ determines the shape of the envelope of the peak primary current $I_{pPK}(\theta) = I_p(T_{ON}, \theta) = Vcs_{REF}(\theta)/R_s$ and, in turn, that of the average inductor current, i.e., the rectified input current $I_{IN}(\theta)$ and ultimately the current $I_{AC}(\theta)$ drawn from the power line. The way the "Current reference generator" block combines the input signals Vc and $V_{MULT}(\theta)$ (and, in case, Q) to produce the reference $Vcs_{REF}(\theta)$ defines the control method.

With the traditional control method [27], which in the following discussion will be referred to as the "QR method", the reference $Vcs_{REF}(\theta)$ is defined by the relationship:

$$Vcs_{REF}(\theta) = K_M Vc V_{MULT}(\theta) \tag{1}$$

where K_M is a constant (multiplier gain, dimensionally 1/V). Being Vc a dc level and $V_{MULT}(\theta)$ a rectified sinusoid, the peaks of the primary current will be enveloped by a sinusoid:

$$I_{pPK}(\theta) = \frac{Vcs_{REF}(\theta)}{Rs} = I_{PPK}\sin\theta, \tag{2}$$

where I_{PPK} is the peak value of the envelope $I_{pPK}(\theta)$. With this method there is an inherent distortion in the input current because the input current flows only during the on-time T_{ON} of the power switch M. Assuming that the turn-on of the power switch is commanded in the instant when the transformer demagnetizes (i.e., assuming $T_V = 0$, see Figure 2 right-hand side), T_{ON} is constant along a line cycle whereas the switching period T is not [27]. The rectified input current is, therefore:

$$I_{IN}(\theta) = I_{PPK}\frac{T_{ON}}{T(\theta)}\sin\theta \tag{3}$$

In [28] an enhanced control method was proposed that produces a reference $Vcs_{REF}(\theta)$ related to the input signals by the relationship:

$$Vcs_{REF}(\theta) = K_M V_C V_{MULT}(\theta)\frac{T(\theta)}{T_{ON}(\theta)} \tag{4}$$

In this way, the peak primary current envelope will not be sinusoidal:

$$I_{pPK}(\theta) = \frac{V_{CSREF}(\theta)}{R_S} = I_{PPK}\frac{T(\theta)}{T_{ON}(\theta)}\sin\theta \qquad (5)$$

but, considering again the approximation $T_V = 0$, the rectified input current will be:

$$I_{IN}(\theta) = I_{PPK}\sin\theta, \qquad (6)$$

so that, in this case, I_{PPK} coincides with the peak value I_{PK} of both $I_{IN}(\theta)$ and $I_{AC}(\theta)$. As previously mentioned, we will refer to this enhanced method as the "EQR method".

3. Input Current Distortion Due to Power Processing: A Closer Look

The simplification $T_V = 0$ used to determine the shape of the rectified input current $I_{in}(\theta)$ described by (3) for the QR method and (6) for the EQR method leads to neglecting the contribution to $I_{IN}(\theta)$ provided by the negative inductor current that flows during this time interval.

The distortion caused by this negative current is discussed in [7], where the analysis has been carried out based on the equivalent circuit depicted in Figure 3. This has been done under the simplifying assumption that the primary current in the instant when M is turned on to start a new switching cycle is zero (Zero-current switching at turn-on, ZCS), as shown in the timing diagrams of Figure 4. Notice that this is equivalent to saying that M is turned on with ZVS (Zero-voltage switching) if $V_{IN} \leq V_R$ and with valley-switching if $V_{IN} > V_R$.

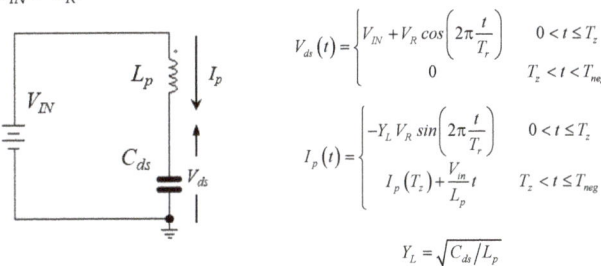

Figure 3. Simplified equivalent circuit of primary side after transformer's demagnetization.

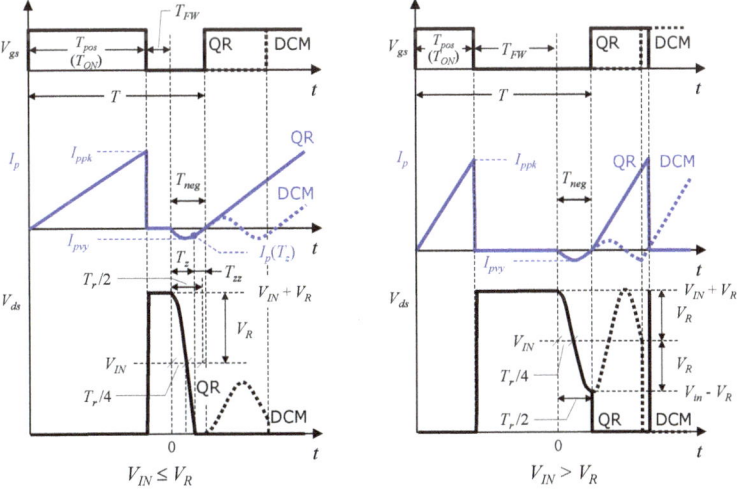

Figure 4. Key waveforms of the circuit in Figure 2 with $V_{IN} \leq V_R$ and zero-current switching (left); with $V_{IN} > V_R$ and valley switching (right). DCM waveforms (dotted lines) are shown for reference.

Assuming that $t = 0$ is the instant when the transformer demagnetizes (i.e., when secondary current zeroes and ringing starts) and t_{ON} the instant when the power switch is turned on, this condition can be labeled as $t_{ON} = T_{neg}$. The results of the analysis, as well as the definition of the relevant timing and electrical quantities, are summarized in Table 1.

Table 1. Timings and primary current characteristics of Hi-PF QR flyback converters assuming that current in the turn-on instant of power switch M is zero ($t_{ON} = T_{neg}$).

Symbol	Definition	Expression
T_{ON}	Duration of ON-time of power switch M	$\frac{L_p I_{ppk}}{V_{IN}}$
T_{FW}	Time needed for secondary current I_s to ramp linearly from I_{spk} down to zero	$\frac{L_s I_{spk}}{V_{out}} = \frac{L_p I_{ppk}}{V_R}$
T	Switching period	$T_{pos} + T_{FW} + T_{neg}$
T_r	Period of drain voltage ringing after demagnetization	$2\pi\sqrt{L_p C_{ds}}$
T_{pos}	Duration of positive portion of primary current I_p.	T_{ON}
T_z	Time needed for drain voltage V_{ds} to fall to zero when $V_{IN} \leq V_R$ or to a minimum when $V_{IN} > V_R$	$\begin{cases} V_{IN} \leq V_R & \frac{T_r}{2}\left(1 - \frac{1}{\pi}\cos^{-1}\frac{V_{IN}}{V_R}\right) \\ V_{IN} > V_R & \frac{T_r}{2} \end{cases}$
T_{zz}	Time needed for primary current I_p to ramp linearly from $I_p(T_z)$ to zero when $V_{IN} \leq V_R$. To be considered equal to zero when $V_{IN} > V_R$	$\begin{cases} V_{IN} \leq V_R & \frac{T_r}{2\pi}\frac{V_R}{V_{IN}}\sqrt{1 - \left(\frac{V_{IN}}{V_R}\right)^2} \\ V_{IN} > V_R & 0 \end{cases}$
T_{neg}	Duration of negative portion of primary current I_p	$\begin{cases} V_{IN} \leq V_R & T_z + T_{zz} \\ V_{IN} > V_R & \frac{T_r}{2} \end{cases}$
Q_{pos}	Positive charge taken from input source during T_{pos} in a switching cycle	$\frac{I_{pkk} T_{pos}}{2}$
Q_{neg}	Negative charge returned to input source during T_{neg} in a switching cycle	$-\frac{C_{ds}}{2}\frac{(V_{IN}+V_R)^2}{V_{IN}}$
$\langle I_p \rangle^*$	Average input current in a switching cycle	$\frac{Q_{pos}+Q_{neg}}{T}$

The ZCS assumption is not always true in practice, because the control circuit that initiates a new switching cycle upon detecting the transformer's demagnetization (ZCD circuit) is not always realized in such a way that the power switch M can be always turned on in the exact instant when the ringing current after demagnetization zeroes.

3.1. Effects of a Nonzero Current at the Turn-on Instant of the Power Switch M

A nonzero current at the turn-on instant will alter the Q_{pos} and/or the Q_{neg} contribution in a switching cycle and this, in turn, will have an impact on the input current shape and, ultimately, on its THD. Both the quantitative and the qualitative effects of this impact have not been analyzed in the existing literature and will be addressed in this section.

The impact is different depending on whether one analyzes the open-loop operation (i.e., with assigned input and output voltages and a profile of the peak primary current $I_{pPK}(\theta)$ having a fixed amplitude) or the closed-loop operation. In this second case, the input voltage is assigned but the amplitude of the profile of the peak primary current $I_{pPK}(\theta)$ is determined by the control loop to deliver the average power demanded by the load in a line cycle with a regulated output voltage or current.

The analysis carried out in this section refers to the open-loop operation.

In the following analysis, the definitions of the quantities considered in Table 1 do not change. To distinguish the quantities related to the $t_{ON} \neq T_{neg}$ case from those related to the $t_{ON} = T_{neg}$ case, the former ones will have an "*" superscript.

It is worth reminding that instantaneous values of all time-varying quantities are considered a function of the phase angle $\theta = 2\pi f_L\, t$ when considering their evolution on a line cycle time scale. Instantaneous values of those quantities that vary within a switching cycle as well are a function of both phase angle and time, being intended that time extends over a single switching cycle, during which the phase angle can be considered constant. To simplify the notation, these dependances will not be explicitly indicated.

We will consider two fundamental cases.

- Case I: $0 < t_{ON} < T_{neg}$ (refer to Figure 5)

In this hypothesis, the initial current $I_p(t_{ON})$ is negative and $T_{ON} > T_{pos}$. Q_{neg} may be affected by the t_{ON} value; considering an open-loop operation, Q_{pos} and T_{pos} will not be affected, in closed-loop operation they will be: a different Q_{neg} value due to a different t_{ON} value needs to be compensated by an opposite change in Q_{pos} so that the input current to the converter is such that the average power delivered to the load in a line cycle does not change. T_{pos} will need to change accordingly.

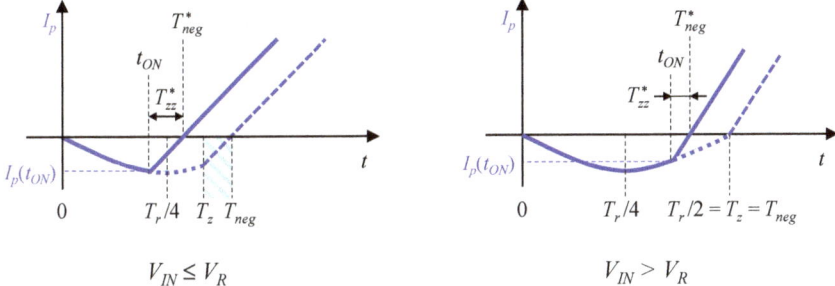

Figure 5. Close-up of primary current with $t_{ON} < T_{neg}$. Dashed lines, which refer to the case $t_{ON} = T_{neg}$, are shown for reference. The diagonally striped area marks the interval where T^*_{neg} and Q^*_{neg} are not affected by t_{ON} position (subsubcase I a2).

We need to distinguish two subcases.

○ Subcase I(a): $V_{IN} \leq V_R$.

We need to distinguish two further subdivisions.

■ Subsubcase I(a1): $0 < t_{ON} \leq T_z$.

In this case, the sinusoidal portion of the negative current will be truncated by the turn-on of the power switch before the drain voltage touches zero. Turn-on will not be exactly ZVS (Zero-voltage switching). The duration of the negative portion of the primary current will be reduced ($T^*_{neg} < T_{neg}$) and so will be Q^*_{neg}.

■ Subsubcase I(a2): $T_z < t_{ON} \leq T_{neg}$.

In this case, the turn-on of the power switch occurs while the primary current, though negative, is already ramping up linearly as if the power switch were turned on at $t = T_z$. Turn-on will still be exactly ZVS. Both T^*_{neg} and Q^*_{neg} will be unaffected ($T^*_{neg} = T_{neg}$, $Q^*_{neg} = Q_{neg}$).

○ Subcase I(b): $V_{IN} > V_R$.

The situation becomes similar to that when $V_{IN} \leq V_R$: current is sinusoidal until $t = t_{ON}$, after that it is a linear ramp. Turn-on will occur before the drain voltage reaches the valley, so valley switching will be lost. Both T^*_{neg} and Q^*_{neg} will be reduced.

The results of this analysis are summarized in Table 2, where T_z is that defined in Table 1 (it is unaffected by t_{ON} and is not shown).

The diagrams in Figure 6 show how T^*_{neg} and Q^*_{neg} vary as a function of the ratio V_{IN}/V_R for different values of t_{ON}. Values are normalized to those for $V_{IN} > V_R$ ($T_r/2$ and $2 V_R C_{ds}$ respectively). The diagrams in Figure 7 show how T^*_{neg} and Q^*_{neg} vary with t_{ON} for different values of the ratio V_{IN}/V_R. Values are normalized in the same manner.

• Case II: $T_{neg} < t_{ON} < T_{neg} + T_r/2$ (refer to Figure 8)

In this case, the initial current $I_p(t_{ON})$ is positive and $T_{ON} < T_{pos}$. Q_{pos} and T_{pos} will be affected, whereas Q_{neg} and T_{neg} will not: Q_{neg} depends on voltages only and not on the power circulating in the converter. Regardless of whether V_{IN} is greater or less than V_R, turn-on occurs on the positive wave of the drain voltage ringing and with the positive current if $t_{ON} < T_r$. However, we need to distinguish two subcases.

○ Subcase II(a): $V_{IN} \leq V_R$.

In this case, due to the charge not transferred from L_p to C_{ds} while the drain voltage is clamped in the interval $T_z \leq t \leq T_{neg}$, the ringing occurring after T_{neg} has a peak amplitude reduced from V_R to V_{IN} in voltage and from $V_R Y_L$ to $V_{IN} Y_L$ in current:

$$V_{ds}(t) = V_{IN}\left[1 - \cos\left(2\pi\frac{t-T_{neg}}{T_r}\right)\right]$$
$$I_p(t) = Y_L V_{IN} \sin\left(2\pi\frac{t-T_{neg}}{T_r}\right) \quad (7)$$

Table 2. Timings and input current characteristics of Hi-PF QR flyback converters assuming current in the turn-on instant of power switch M is negative ($t_{ON} < T_{neg}$).

Symbol	Definition	Expression
T^*_{pos}	Duration of Positive Portion of Primary Current I_p.	$\frac{L_p I_{ppk}}{V_{IN}}$
T^*_{ON}	Duration of ON-time of power switch M	$T^*_{pos} + T^*_{neg} - t_{ON}$
T^*_{FW}	Time needed for secondary current I_s to ramp linearly from I_{spk} down to zero	$\frac{L_s I_{spk}}{V_{out}} = \frac{L_p I_{ppk}}{V_R}$
T^*	Switching period	$T^*_{pos} + T^*_{FW} + T^*_{neg}$
T^*_{zz}	Time needed for primary current I_p to ramp linearly to zero starting from $I_p(t_{ON})$	$\frac{1}{2\pi} T_r \frac{V_R}{V_{IN}} \sin\left(2\pi\frac{t_{ON}}{T_r}\right)$
T^*_{neg}	Duration of negative portion of primary current I_p	$V_{IN} \leq V_R,\ 0 \leq t_{ON} < T_z$: $t_{ON} + T^*_{zz}$; $V_{IN} \leq V_R,\ T_z \leq t_{ON} \leq T_{neg}$: T_{neg}; $V_{IN} > V_R$: $t_{ON} + T^*_{zz}$
Q^*_{pos}	Positive charge taken from input source during T_{pos} in a switching cycle	$\frac{1}{2} I_{ppk} T^*_{pos}$
Q^*_{neg}	Negative charge returned to input source during T^*_{neg} in a switching cycle	$V_{IN} \leq V_R,\ 0 \leq t_{ON} < T_z$: $V_R C_{ds}\left[\cos\left(2\pi\frac{t_{ON}}{T_r}\right) + -\frac{1}{2}\frac{V_R}{V_{IN}}\sin^2\left(2\pi\frac{t_{ON}}{T_r}\right) - 1\right]$; $V_{IN} \leq V_R,\ T_z \leq t_{ON} \leq T_{neg}$: Q_{neg}; $V_{IN} > V_R$: $V_R C_{ds}\left[\cos\left(2\pi\frac{t_{ON}}{T_r}\right) + -\frac{1}{2}\frac{V_R}{V_{IN}}\sin^2\left(2\pi\frac{t_{ON}}{T_r}\right) - 1\right]$
$I_p(t_{ON})$	Primary current at turn-on instant	$V_{IN} \leq V_R,\ 0 \leq t_{ON} < T_z$: $-Y_L V_R \sin\left(2\pi\frac{t_{ON}}{T_r}\right)$; $V_{IN} \leq V_R,\ T_z \leq t_{ON} \leq T_{neg}$: $\frac{V_{IN}}{L_p}(t_{ON} - T_{neg})$; $V_{IN} > V_R$: $-Y_L V_R \sin\left(2\pi\frac{t_{ON}}{T_r}\right)$
$\langle I_p \rangle^*$	Average input current in a switching cycle	$\frac{Q^*_{pos} + Q^*_{neg}}{T^*}$

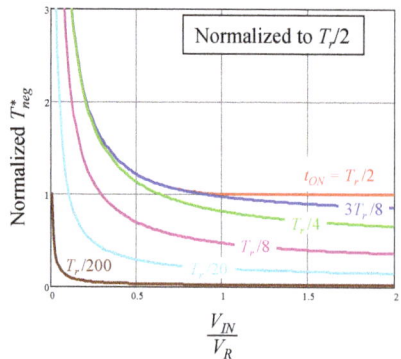 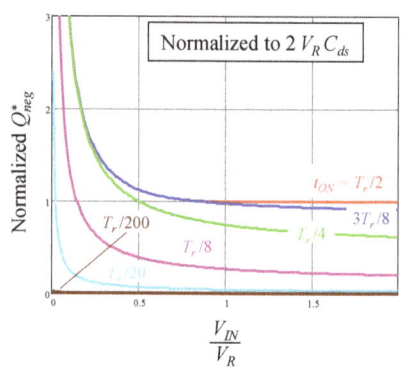

Figure 6. Normalized values of T^*_{neg} and Q^*_{neg} as a function of the V_{IN}/V_R ratio for different turn-on instants t_{ON}, with $t_{ON} \leq T_{neg}$.

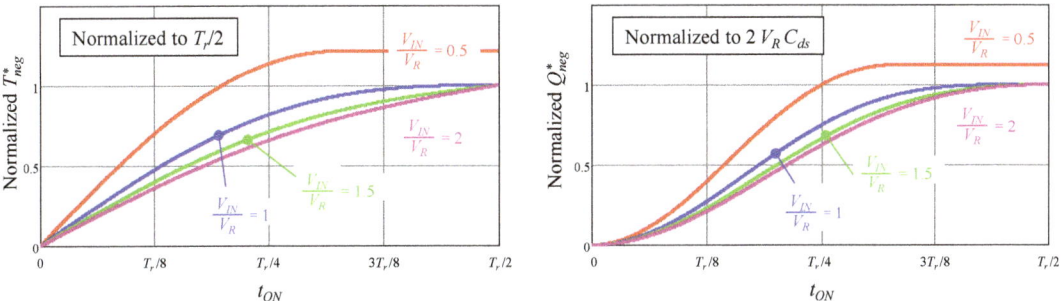

Figure 7. Normalized values of T^*_{neg} and Q^*_{neg} as a function of the turn-on instant t_{ON}, with $t_{ON} \leq T_{neg}$, for different V_{IN}/V_R ratios.

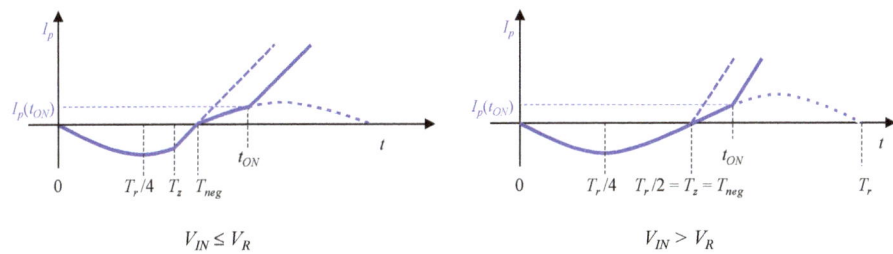

Figure 8. Close-up of primary current with $t_{ON} > T_{neg}$. Dashed lines, which refer to the case $t_{ON} = T_{neg}$, are shown for reference. Dotted lines show primary current ringing continuation.

- Subcase II(b): $V_{IN} \leq V_R$.

In this case, the exchange of electric charge between L_p and C_{ds} is unaffected, and there is no change in the ringing occurring after $T_r/2$.

The results of this analysis are summarized in Table 3, where T_z and T_{zz} (which are those defined in Table 1) are not shown because not relevant and unaffected by t_{ON}.

The diagrams in Figure 9 show how T^*_{pos} and Q^*_{pos} vary as a function of the ratio V_{IN}/V_R for different values of t_{ON}. Values are normalized to those of T_{pos} and Q_{pos} when $t_{ON} = T_{neg}$. The diagrams in Figure 10 show how T^*_{neg} and Q^*_{neg} vary with t_{ON} for different values of the ratio V_{IN}/V_R. Values are normalized in the same manner.

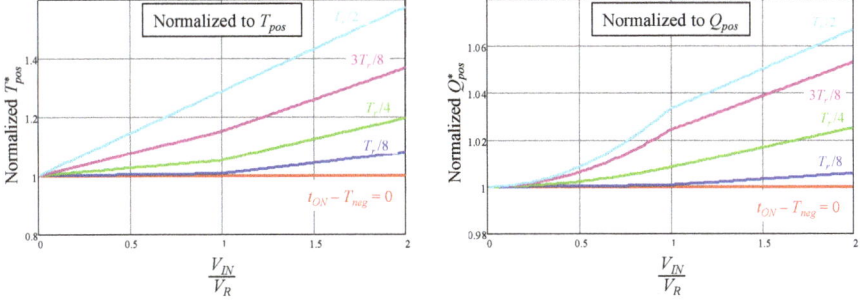

Figure 9. Normalized values of T^*_{pos} and Q^*_{pos} as a function of the V_{IN}/V_R ratio for different turn-on instants t_{ON}, with $t_{ON} \geq T_{neg}$.

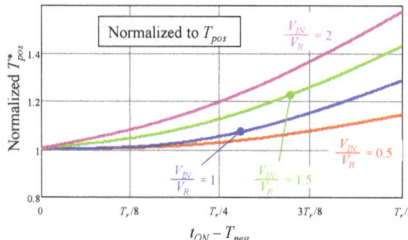
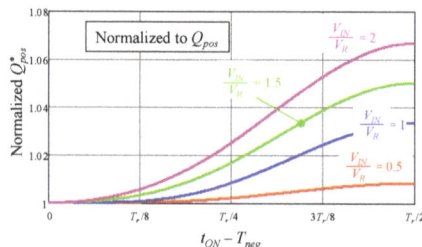

Figure 10. Normalized values of T^*_{pos} and Q^*_{pos} as a function of the turn-on instant t_{ON}, with $t_{ON} \geq T_{neg}$, for different V_{IN}/V_R ratios.

Table 3. Timings and input current characteristics of Hi-PF QR flyback converters assuming current in the turn-on instant of power switch M is positive ($t_{ON} > T_{neg}$).

Symbol	Definition	Expression
T^*_{pos}	Duration of positive portion of primary current I_p.	$T^*_{ON} + t_{ON}$
T^*_{ON}	Duration of ON-time of power switch M	$\begin{cases} V_{IN} \leq V_R \\ V_{IN} > V_R \end{cases}$ $\frac{L_p}{V_{IN}} I_{ppk} - \frac{1}{2\pi} T_r \sin\left(2\pi \frac{t_{ON}}{T_r}\right)$ $\frac{L_p}{V_{IN}} I_{ppk} - \frac{1}{2\pi} T_r \frac{V_R}{V_{IN}} \sin\left(2\pi \frac{t_{ON}}{T_r}\right)$
T^*_{FW}	Time needed for secondary current I_s to ramp linearly from I_{spk} down to zero	$\frac{L_s I_{spk}}{V_{out}} = \frac{L_p I_{ppk}}{V_R}$
T^*	Switching period	$T^*_{pos} + T^*_{FW} + T^*_{neg}$
T^*_{neg}	Duration of negative portion of primary current I_p	T_{neg}
Q^*_{neg}	Negative charge returned to input source during T^*_{neg} in a switching cycle	Q_{neg}
Q^*_{pos}	Positive charge taken from input source during T^*_{pos} in a switching cycle	$\begin{cases} V_{IN} \leq V_R \\ V_{IN} > V_R \end{cases}$ $\frac{1}{2}\left[I_{ppk} + V_{IN} Y_L \sin\left(2\pi \frac{t_{ON}}{T_r}\right)\right] T^*_{ON}$ $\frac{1}{2}\left[I_{ppk} + V_R Y_L \sin\left(2\pi \frac{t_{ON}}{T_r}\right)\right] T^*_{ON}$
$I_p(t_{ON})$	Primary current at turn-on instant	$\begin{cases} V_{IN} \leq V_R \\ V_{IN} > V_R \end{cases}$ $Y_L V_{IN} \sin\left(2\pi \frac{t_{ON}}{T_r}\right)$ $Y_L V_R \sin\left(2\pi \frac{t_{ON}}{T_r}\right)$
$\langle I_p \rangle^*$	Average input current in a switching cycle	$\frac{Q^*_{pos} + Q^*_{neg}}{T^*}$

3.2. Comments on Previous Analysis

It is worth reminding that this analysis contains simplifications that impact the quantitative aspect. Especially noteworthy is the one concerning the C_{oss} of the power switch M, a strongly nonlinear capacitance that in the latest MOSFET generations increases dramatically (a hundred times or more) when the drain-source voltage falls below few ten volts. This capacitance has been considered constant or, at least, not significantly impacting the overall C_{ds}.

Another simplification is that the ringing has been assumed to be undamped. Actually, the ringing current flowing through the primary winding of the transformer encounters the ac resistance of that winding at the ringing frequency (typically, some hundred kHz). Because of skin and proximity effects, and depending on the construction of the transformer, its value may be even significantly high and provide significant damping of the ringing. In the context of the present analysis, the most significant consequence of the damping is that the amplitude of the ringing, even considering the first valley, will be lower than V_R. Therefore, the valley of the drain voltage will touch zero at an input voltage lower than V_R.

Another simplification concerns what happens when the ac line voltage $V_{AC}(\theta)$ approaches zero.

The present analysis points out the existence of a time interval around zero crossings of $V_{AC}(\theta)$ (often termed dead zone) where the input current to the converter $I_{AC}(\theta) = 0$, although $V_{AC}(\theta) \neq 0$, originating the so-called crossover distortion. This happens when the peak current in a switching cycle becomes so small that $Q_{pos} < |Q_{neg}|$ and the rectified input current $I_{IN}(\theta)$ becomes negative. The physical interpretation of being $I_{IN}(\theta) < 0$ and

$I_{AC}(\theta) = 0$ around the zero-crossings is: when $I_{IN}(\theta) < 0$ it actually charges back the input capacitor (C_{in} in Figure 1) so that $V_{IN}(\theta)$ becomes larger than $V_{AC}(\theta)$, the input bridge is reverse-biased and, consequently, $I_{AC}(\theta)$ is zero.

The situation is even more complicated because of the residual voltage across the input capacitance C_{in} and the possible interactions with the input EMI filter due to the drastic reduction of the switching frequency that occurs near the zero-crossings. Additionally, there is another dead zone around the line voltage zero-crossings where no primary-to-secondary energy transfer takes place that interacts with that determined by the average inductor current $I_{IN}(\theta)$ being negative. All these aspects are expanded in [7].

For the completeness of the analysis, it is worth mentioning another minor effect that makes the task of an accurate description of the behavior near the zero crossings even tougher.

As reported in [7], the previously mentioned region around the line voltage zero crossings where there is no input-to-output energy transfer occurs when the inductor peak current is so low that the energy stored in L_p during the on-time T_{ON} of the power switch M is not enough to charge C_{ds} up to $V_{IN}(\theta) + V_R$ when the power switch turns off.

As illustrated in Figure 11, the drain waveform is a sinusoidal arc going from zero to a peak value $V_Y < V_R$ and then back to zero. The situation is significantly different as compared to that considered during the previous analysis and shown in Figure 4: V_Y is not large enough for the voltage on the secondary side of the transformer to forward bias the rectifier and have current circulating, thus the T_{FW} interval disappears and no energy is delivered to the output.

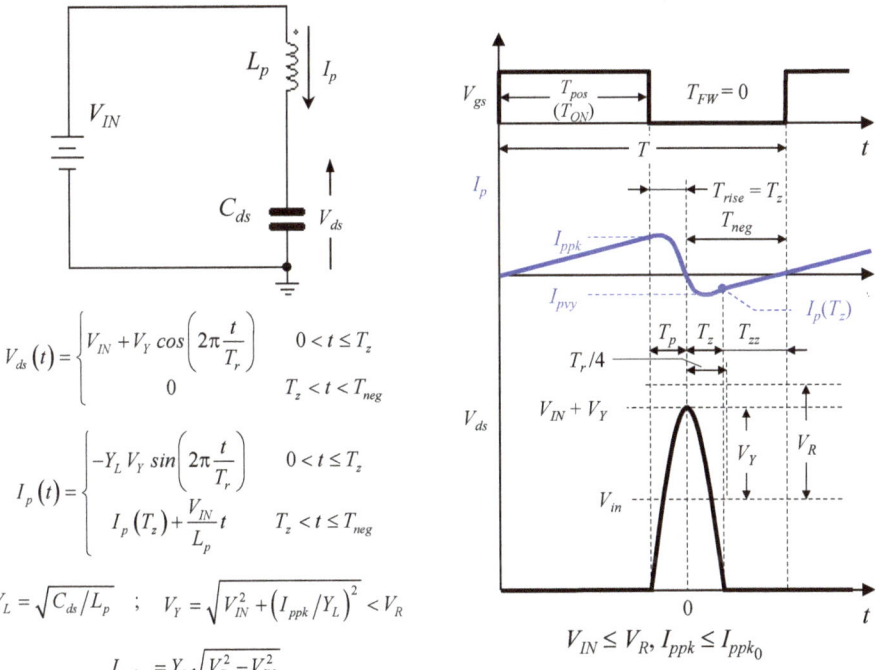

Figure 11. Simplified equivalent circuit after power switch turn-off in the no input-to-output energy transfer region around the line voltage zero-crossings and relevant key waveforms.

Assuming as $t = 0$ the instant when the drain voltage peaks, the interval T_z needed for it to fall to zero is essentially the same as that needed to reach the peak after turn-off. Both the rise and fall times are then in the range of $T_r/4$. The rise time of the drain voltage, which has always been assumed to be negligible, in this case, is not so short but the duration of the switching period is largely dominated by the interval $T_{zz} + T_{pos}$ because of the very low input voltage, then its effect on the switching frequency can still be neglected.

The equations shown in Figure 2 and that have been used to derive those in Tables 1–3 are no longer valid and should be modified by substituting V_R with V_Y.

3.3. Quantitative Aspects of a Nonzero Current at the Turn-on Instant of the Power Switch M

To provide a quantitative idea of the impact of a nonzero current at the turn-on instant in the closed-loop operation, the analysis done so far will be applied to a pair of exemplary cases.

The first exemplary case is a Hi-PF QR flyback converter whose main electrical specification is detailed in Table 4 and that is controlled with the traditional QR method, where the peak primary current is enveloped by a rectified sinusoid as expressed by (2).

Table 4. Main characteristics of Hi-PF QR flyback converter with the QR control method used as a first reference.

Parameter	Symbol	Value	Unit
Line voltage range	V_{ACmin}–V_{ACmax}	90–265	Vac
Line frequency range	f_L	47–63	Hz
Regulated output voltage	Vout	48	V
Rated dc output current	Iout	730	mA
Expected full-load efficiency	η	90	%
Reflected voltage	V_R	180	V
Transformer primary inductance	L_p	550	µH
Output capacitance	C_{out}	2000	µF
Input capacitance	C_{in}	220	nF
Drain node total capacitance	C_{ds}	140	pF
Switching frequency range (@ line voltage peaks)	f_{swmin}–f_{swmax}	≈64–150	kHz

This control method provides a rectified input current to the converter, $I_{IN}(\theta)$, which is given by the sequence of $\langle I_p \rangle$ along each line half-cycle, that is expressed by:

$$I_{IN}(\theta) = \begin{cases} \frac{1}{2}I_{PPK}\frac{T_{pos}(\theta)}{T(\theta)}\sin\theta - \frac{2}{T(\theta)}V_R C_{ds} & V_{IN}(\theta) > V_R \\ \frac{1}{2}I_{PPK}\frac{T_{pos}(\theta)}{T(\theta)}\sin\theta - \frac{1}{2T(\theta)}\frac{[V_{IN}(\theta)+V_R]^2}{V_{IN}(\theta)}C_{ds} & V_{IN}(\theta) \leq V_R \end{cases} \quad (8)$$

In this converter the turn-on instant t_{ON} has been swept in the interval $0 \leq t_{ON} \leq T_r$, considering operation at low line voltage (115 Vac) and high line voltage (230 Vac) at full load. The results are shown in the diagram in Figure 12.

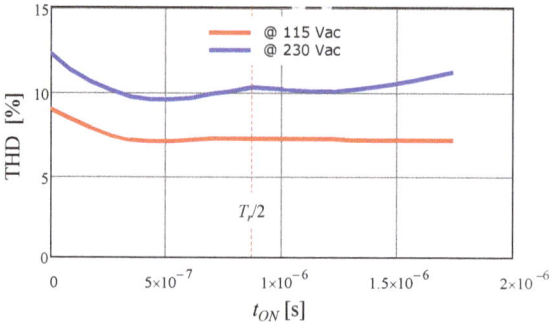

Figure 12. Total harmonic distortion (THD) of the input current $I_{AC}(\theta)$ of the converter specified in Table 4, controlled with the QR control method, upon varying the turn-on instant t_{ON} of the power switch in the interval $0 \leq t_{ON} \leq T_r$.

Notice that in a large region around $t_{ON} = T_r/2$, the one of greater practical interest, the distortion is nearly independent of t_{ON}, especially at the low line. Based on this observation, it is possible to conclude that with the QR control method the distortion of the input current is essentially insensitive to the way the ZCD circuit is realized and to the statistical spread of its parameters.

The second exemplary case is a Hi-PF QR flyback converter whose main electrical specification is detailed in Table 5 and that is controlled with the EQR method, where the envelope of the peak primary current is given by (5).

Table 5. Main characteristics of Hi-PF QR flyback converter with the EQR control method used as a second reference.

Parameter	Symbol	Value	Unit
Line voltage range	$V_{ACmin} - V_{ACmax}$	90–265	Vac
Line frequency range	f_L	47–63	Hz
Rated output voltage (14 LED string @ 100% load)	Vout	48	V
Regulated dc output current	Iout	730	mA
Expected full-load efficiency	η	90	%
Reflected voltage	V_R	120	V
Transformer primary inductance	L_p	500	µH
Output capacitance	C_{out}	1360	µF
Input capacitance	C_{in}	470	nF
Drain node total capacitance	C_{ds}	220	pF
Switching frequency range (@ line voltage peaks)	$f_{swmin} - f_{swmax}$	≈44–88	kHz

This control method provides a rectified input current to the converter, $I_{IN}(\theta)$, expressed by:

$$I_{IN}(\theta) = \begin{cases} \frac{1}{2} I_{PPK} \frac{T_{pos}(\theta)}{T_{ON}(\theta)} \sin\theta - \frac{2}{T(\theta)} V_R C_{ds} & V_{IN}(\theta) > V_R \\ \frac{1}{2} I_{PPK} \frac{T_{pos}(\theta)}{T_{ON}(\theta)} \sin\theta - \frac{1}{2T(\theta)} \frac{[V_{IN}(\theta) + V_R]^2}{V_{IN}(\theta)} C_{ds} & V_{IN}(\theta) \leq V_R \end{cases} \quad (9)$$

In this converter too the turn-on instant t_{ON} has been swept in the interval $0 \leq t_{ON} \leq T_r$ considering operation at 115 Vac and 230 Vac at full load. The results are shown in the diagram in Figure 13.

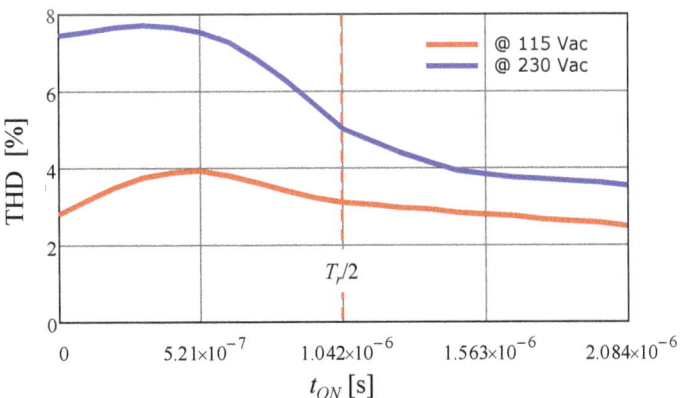

Figure 13. Total harmonic distortion (THD) of the input current $I_{AC}(\theta)$ of the converter specified in Table 5, controlled with the EQR control method, upon varying the turn-on instant t_{ON} of the power switch in the interval $0 \leq t_{ON} \leq T_r$.

Notice that in this case in the region around $t_{ON} = T_r/2$, at the low line the distortion is a little dependent on t_{ON}, whereas at the high line the dependence is significantly higher.

In both cases, the distortion becomes smaller as t_{ON} is delayed. Based on this observation, it is possible to conclude that with the EQR control method the distortion of the input current, though lower as compared to the QR method, is more sensitive to t_{ON} and, then, to the way the ZCD circuit is realized and to the statistical spread of its parameters.

Notice that in the positive terms in (5), the ratio $T_{pos}(\theta)/T_{ON}(\theta)$ expresses the effect that the current in the turn-on instant of the power switch M has on the shape of the input current. In fact, in the case $T_{pos}(\theta) \neq T_{ON}(\theta)$, since in general, they are not proportional to one another, the positive term contains a distortion term that adds up to the distortion caused by the negative one. This explains why the EQR method is more sensitive to t_{ON} and to the implementation of the ZCD circuit.

3.4. Notes on the THD Calculation Method

The THD of $I_{AC}(\theta)$ vs. t_{ON} plots of Figures 12 and 13 have been derived determining the expressions of $I_{IN}(\theta)$, Equation (8) for the QR method and Equation (9) for the EQR method, and then extending them over the interval $(0, 2\pi)$. In these equations I_{PPK}, which is related to the power delivered by the converter, is the unknown parameter, t_{ON} is the independent variable, the others are those in Tables 2 and 3.

Notice that the quantity I_{PPK} in these tables is given by Equation (2) for the QR method and Equation (5) for the EQR method. To prevent the singularity when $V_{IN} = 0$ in the expressions related to $V_{IN} \leq V_R$ where V_{IN} appears at the denominator, a voltage drop $V_F = 0.7$ V across the body diode of the power switch has been assumed. Hence, in all formulas in Table 2 where $V_{IN} \leq V_R$ it is actually $V_{IN} = V_{PK} \sin \theta + V_F$.

In closed-loop operation, for a given V_{AC}, the average value of the product $V_{IN}(\theta) \cdot I_{IN}(\theta)$ over a line half-cycle (i.e., the integral of the product in $(0, \pi)$ divided by π) must be equal to the dc input power to the converter $Pin = Vout\ Iout/\eta$. This equation is solved by iteration for the unknown parameter I_{PPK} with a given value of t_{ON}, thus completely defining Equations (8) and (9).

As previously said, the Fourier coefficients (i.e., the peak amplitudes of each harmonic) of $I_{AC}(\theta)$ are computed by extending Equations (8) and (9) over the interval $(0, 2\pi)$. Being $I_{AC}(\theta)$ an odd function, there are only sine terms. Because of its rotational symmetry $(I_{AC}(\theta + p) = -I_{AC}(\theta))$ there are only odd harmonics.

The THD is computed as the ratio between the square root of the sum of squares of the peak amplitudes of the higher order harmonics (from 3rd up to 39th) and the peak amplitude of the fundamental one.

Finally, all these calculations are repeated sweeping t_{ON} from 0 to T_r. This has been done with the help of Mathcad®, engineering math software.

4. Impact of ZCD Circuit Operation on Input Current Shaping

The impact of a nonzero current as the power switch M turns on, causing $T_{pos}(\theta) \neq T_{ON}(\theta)$, will be now analyzed more in detail with reference to some typical implementations of the ZCD circuit that determines the turn-on instant t_{ON} of the power switch M. Some results, referred to the exemplary converter specified in Table 5 and controlled with the EQR method will be presented to provide the reader with some quantitative information. No investigation will be done on the converter specified in Table 4 and controlled with the QR method, since the previous analysis has shown its substantial insensitivity to the position of t_{ON} and, then, to the operation of the ZCD circuit.

The calculation method used to obtain the results that are shown in the following sections is the same as that described in Section 3.4.

4.1. Optimal ZCD Circuit

An optimal ZCD circuit ensures that the turn-on of the power switch always occurs with zero initial current ($t_{ON} = T_{neg}$), so that it is always $T_{pos} = T_{ON}$ and, as stated by Equation (9), no distortion is introduced in the positive terms all over the V_{IN} range:

$$I_{IN}(\theta) = \begin{cases} \frac{1}{2}I_{PPK}\sin\theta - \frac{2}{T(\theta)}V_R C_{ds} & V_{IN}(\theta) > V_R \\ \frac{1}{2}I_{PPK}\sin\theta - \frac{1}{2T(\theta)}\frac{[V_{IN}(\theta)+V_R]^2}{V_{IN}(\theta)}C_{ds} & V_{IN}(\theta) \leq V_R \end{cases} \quad (10)$$

The principle circuit shown in Figure 14 along with its key waveforms may fulfill this task. The auxiliary winding L_{aux} is coupled to the primary winding of the transformer in such a way that its voltage V_{aux} and the drain voltage V_{ds} are in-phase. More specifically, V_{aux} is a replica of the drain voltage V_{ds} scaled down by the turn ratio and centered on zero.

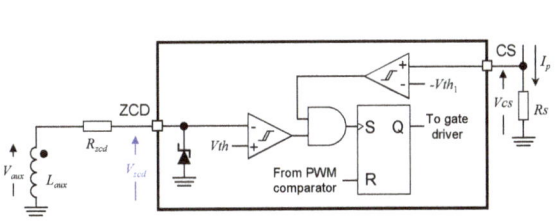
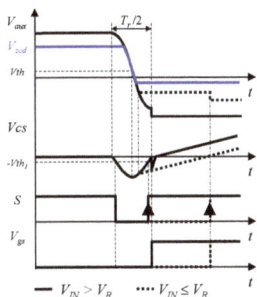

Figure 14. Principle schematic of an optimal ZCD circuit and relevant key waveforms.

V_{aux}, which is negative during the on-time of the power switch M, is positive during the off-time, as long as the current circulates on the secondary winding; when this current zeroes (demagnetization), V_{ds} starts ringing with a negative-going sinusoidal arc and the same falling arc appears on V_{aux}.

L_{aux} is coupled to the ZCD pin of the control IC via the resistor R_{zcd}: since the voltage V_{zcd} on the ZCD pin is top and bottom clamped, R_{zcd} limits the current sunk/sourced by the clamps.

A comparator (ZCD comparator) with the noninverting input referred to a slightly positive threshold Vth (e.g., 100 mV) senses V_{zcd} on its inverting input. Another comparator (CS comparator), whose inverting input is referred to as a negative threshold $-Vth_1$ very close to zero (e.g., -20 mV) senses the voltage on the current sense input V_{CS} on its noninverting input. The PWM latch is edge-sensitive and its set input S is driven by the AND gate that receives the outputs of the two comparators. With this circuit arrangement, the output Q of the PWM latch goes high causing the power switch M to turn on both conditions, $V_{zcd} < Vth$ and $V_{CS} > -Vth_1$, are met.

When the secondary current zeroes, V_{aux} collapses and, as it goes below the upper clamp value, also V_{zcd} starts collapsing. As V_{zcd} falls below Vth the output of the ZCD comparator goes high. Being Vth close to zero, this occurs about $T_r/4$ after the secondary current zeroes. The primary current I_p is ringing too (in quadrature to V_{aux}), so in that instant, I_p is close to its negative peak, it is $V_{CS} < -Vth_1$ and the output of the CS comparator is low. As long as I_p is negative and it is $V_{CS} < -Vth_1$, the output of the CS comparator stays low. Only when V_{CS} exceeds $-Vth_1$ (either because of ringing when $V_{IN} > V_R$, or because I_p is ramping up linearly when $V_{IN} \leq V_R$), the output of the CS comparator and the output of the AND gate go high too. The PWM latch is then set, its output Q goes high and turns on the gate driver and the power switch M, starting a new switching cycle.

The diagrams of Figure 15 provide some exemplary quantitative results for the converter specified in Table 5. The diagrams on the left-hand side show the shape of the input current to the converter $I_{AC}(\theta)$ (in red) along with a black sinusoid for reference and, below, its harmonic contents at full load and $Vac = 115$ V. The diagrams on the right-hand side show the same at $Vac = 230$ V.

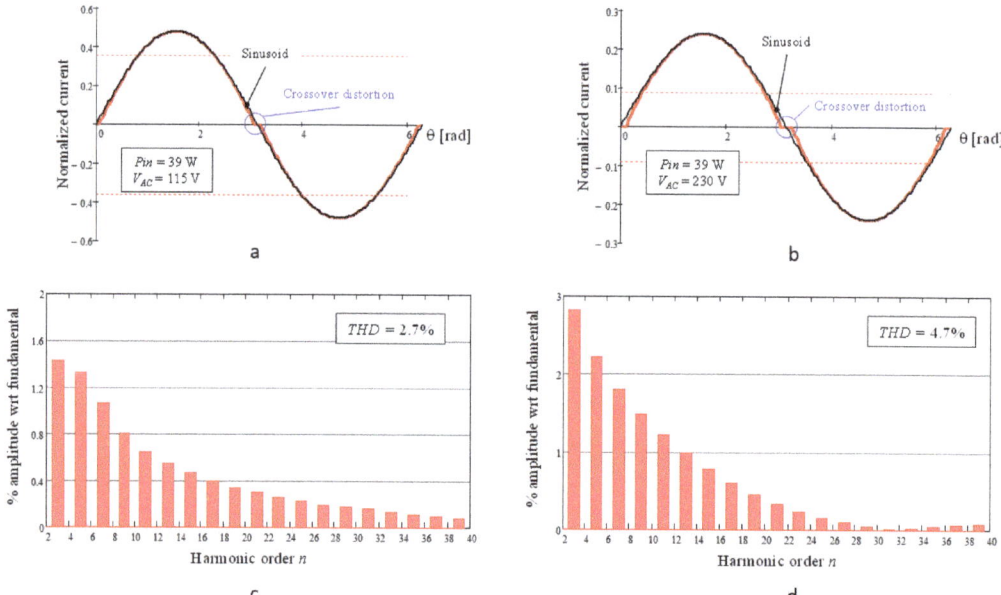

Figure 15. Optimal ZCD circuit: input current shape (**a,b**) and its harmonic content (**c,d**) for converter specified in Table 5 at 115 Vac (**a,c**) and 230 Vac (**b,d**).

The horizontal red dotted lines in the upper diagrams mark the points where $|V_{IN}| = V_R$, i.e., the transition from the region ($|V_{IN}| > V_R$), where the negative charge Q_{neg} depends on V_R only, to the region ($|V_{IN}| < V_R$) of $I_{IN}(\theta)$ and then, of $I_{AC}(\theta)$ where Q_{neg} depends on V_{IN} too.

Note that the shape of $I_{AC}(\theta)$ shows the crossover distortion, highlighted by the blue circle, i.e., the dead zone corresponding to a negative $I_{IN}(\theta)$ around the zero crossings of the instantaneous line voltage $V_{AC}(\theta)$, which makes $I_{AC}(\theta) = 0$, although $V_{AC}(\theta) \neq 0$ as previously explained.

The dead zone in $I_{AC}(\theta)$ predicted by (10) lies in the interval $-3.2° < \theta < 3.2°$ at 115 Vac and in the interval $-5.8° < \theta < 5.8°$ at 230 Vac. As discussed in Section 3.2, the accuracy of the model Equation (10) around line voltage zero-crossings is impaired by the existence of other distortion causes (above all else the input capacitor C_{in}). Therefore, these data on the dead zone amplitude are ballpark figures that can be used only for comparison with other ZCD circuits by isolating their contribution alone.

4.2. Differentiator-Based ZCD Circuit

The principle schematic of this circuit and its key waveforms are shown in Figure 16. Both the external circuit connected to the ZCD pin and its operation are exactly the same as with the optimal ZCD circuit, except that in this case the pin is directly connected to L_{aux}.

Internally, the voltage on the ZCD pin is unclamped; as to the differentiator, it is assumed that it is $R_d C_d \ll T_r/2$. The current I_d (and, then, the voltage $V_d = I_d R_d$) is zero as long as V_{aux} is on either level and is nonzero when V_{aux} transitions from one level to the other. The voltage V_d is sensed by a comparator whose inverting input is referred to a negative threshold -Vth close to zero, e.g., −100 mV.

When the secondary current zeroes and the negative-going edge of V_{aux} starts, the voltage V_d applied to the noninverting input of the comparator becomes much negative, thus its output goes low. When the negative-going edge ends I_d zeroes and so does V_d too: as it exceeds -Vth the output of the comparator has a low-to-high transition. The PWM latch, edge-sensitive, is then set, its output Q goes high turning on the gate driver and the power switch M and starting a new switching cycle.

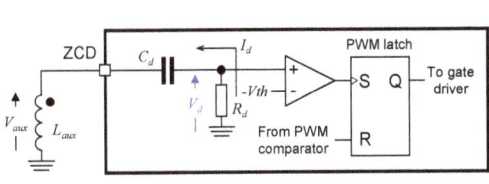
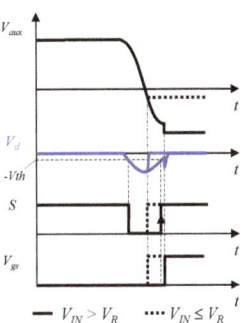

Figure 16. Principle schematic of a differentiator-based ZCD circuit and relevant key waveforms.

With this circuit, if $V_{IN} > V_R$ the turn-on of the power switch M occurs on the valley of the V_{ds} ringing ($t_{ON} = T_r/2$, when its derivative is zero): zero derivative means zero ringing currents and, then, zero initial current and $T_{pos} = T_{ON}$. No distortion is associated with the corresponding positive term in Equation (9). When $V_{IN} \leq V_R$ the turn-on is commanded as the drain voltage touches zero, i.e., $t_{ON} = T_z$ given in Table 1 when the ringing current is still negative. In this case, based on Table 2, T_{ON} is expressed as:

$$T_{ON} = T_{pos} + T_{zz} \tag{11}$$

and (5), neglecting the contribution of the ringing current as previously stated, becomes:

$$I_{IN}(\theta) = \begin{cases} \frac{1}{2}I_{PPK}\sin\theta - \frac{2}{T(\theta)}V_R C_{ds} & V_{IN}(\theta) > V_R \\ \frac{1}{2}I_{PPK}\sin\theta - \frac{1}{2}I_{PPK}\frac{T_{zz}(\theta)}{T_{ON}(\theta)}\sin\theta - \frac{1}{2T(\theta)}\frac{[V_{IN}(\theta)+V_R]^2}{V_{IN}(\theta)}C_{ds} & V_{IN}(\theta) \leq V_R \end{cases} \tag{12}$$

As a conclusion, the differentiator-based ZCD circuit does not introduce any distortion as long as $V_{IN} > V_R$ (condition fulfilled along most of the rectified sinusoid at the high line, e.g., with the European mains); conversely, it introduces a distortion term when $V_{IN} \leq V_R$, a condition that is fulfilled along most of the rectified sinusoid at the low line, i.e., with US or Japan mains). The amplitude of this distortion is related to the ratio T_{zz}/T_{ON} and a Fourier analysis of Equation (12) shows that the distortion term creates a component at the fundamental frequency and odd harmonics, all in phase opposition to the fundamental component.

With reference again to the converter specified in Table 5, the diagrams of Figure 17 provide the same exemplary quantitative results as those shown in Figure 15 with the optimal ZCD circuit. Note that the harmonic contents have a distribution not too different from that provided by the optimal ZCD circuit, but with a slightly larger amplitude. This results in a slightly higher THD: +1% at 115 Vac and +1.3% at 230 Vac.

In this case the crossover distortion due to the positive term in Equation (12) becoming smaller than the negative term is slightly wider: the dead zone in $I_{AC}(\theta)$ occurs in the interval $-3.4° < \theta < 3.4°$ at 115 Vac and in $-6.7° < \theta < 6.7°$ at 230 Vac. The external circuit connected to the ZCD pin and its operation are the same as the optimal ZCD circuit. Internally, the voltage V_{zcd} on the ZCD pin is top and bottom clamped and there is the same ZCD comparator with the noninverting input referred to as a slightly positive threshold Vth (e.g., 100 mV) that senses V_{zcd} on its inverting input as seen in the optimal ZCD circuit. The output X of the ZCD comparator goes through a delay block T_d, ideally tuned to slightly less than $T_r/4$, after that it reaches the set input of the edge-sensitive PWM latch.

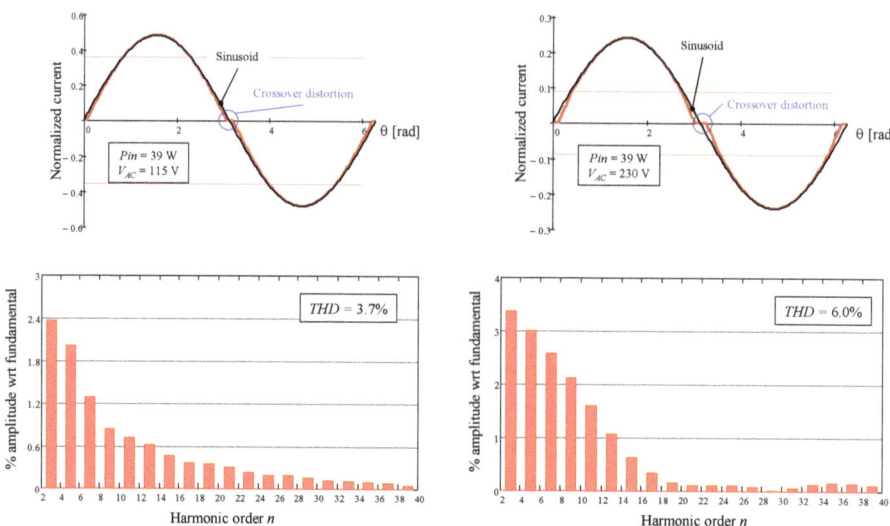

Figure 17. Differentiator-based ZCD circuit: input current shape (upper) and its harmonic content (lower) for converter specified in Table 1 at 115 Vac (left) and 230 Vac (right).

4.3. Comparator-plus-Delay ZCD Circuit

The principle schematic of this circuit and its key waveforms are shown in Figure 18.

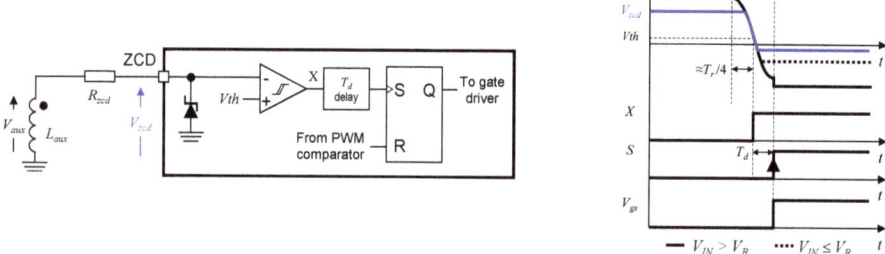

Figure 18. Principle schematic of a comparator-plus-delay ZCD circuit and relevant key waveforms.

When the secondary current zeroes, V_{aux} collapses and, as it goes below the upper clamp value, V_{zcd} also starts collapsing. As V_{zcd} falls below Vth the output X of the comparator has a low-to-high transition. After a delay T_d the PWM latch is set, its output Q goes high and turns on the gate driver and the power switch M, starting a new switching cycle.

Note that, being Vth close to zero, the negative edge is detected about $T_r/4$ after the secondary current zeroes. Note also that the resistor R_{zcd}, along with the parasitic capacitance of the internal clamp plus some external stray contributors (which are anyhow well defined once the layout of the external circuit is defined), form an RC low-pass filter that delays V_{zcd} with respect to V_{aux}. This delay adds up to T_d and can be fine-tuned by adjusting R_{zcd} (or even adding a small external capacitor between the ZCD pin and ground) so that the overall delay equals $T_r/2$ and it is $t_{ON} = T_r/2$ both with $V_{IN} > V_R$ and $V_{IN} \leq V_R$.

With this circuit, therefore, when $V_{IN} > V_R$ turn-on occurs on the valley of the V_{ds} ringing, the initial current is zero, $T_{pos} = T_{ON}$, and no distortion is introduced. When $V_{IN} \leq V_R$ the ringing current at $t = T_r/2$ is ramping up linearly but is still negative. In this case, based on Table 2, T_{ON} is given by:

$$T_{ON} = T_{pos} + T_{neg} - \frac{T_r}{2} \quad (13)$$

and (9) becomes:

$$I_{IN}(\theta) = \begin{cases} \frac{1}{2}I_{PPK}\sin\theta - \frac{2}{T(\theta)}V_R C_{ds} & V_{IN}(\theta) > V_R \\ \frac{1}{2}I_{PPK}\sin\theta - \frac{1}{2}I_{PPK}\frac{T_{neg}(\theta)-\frac{T_r}{2}}{T_{ON}(\theta)}\sin\theta - \frac{1}{2T(\theta)}\frac{[V_{IN}(\theta)+V_R]^2}{V_{IN}(\theta)}C_{ds} & V_{IN}(\theta) \leq V_R \end{cases} \quad (14)$$

Once more with reference to the converter specified in Table 5, the diagrams of Figure 19 provide the same results as those shown in Figures 15 and 17, with the comparator-plus-delay ZCD circuit. Note that the harmonic contents is lower in amplitude as compared to that of the differentiator-based ZCD circuit and quite close to that of the optimal ZCD circuit: the THD values are only 0.4% at 115 Vac and 0.3% at 230 Vac larger.

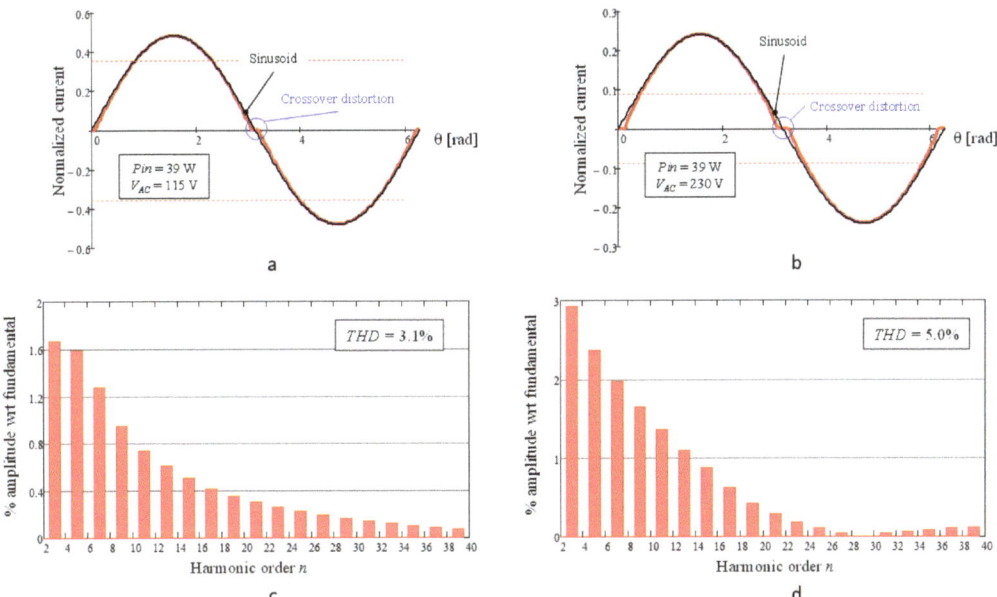

Figure 19. Comparator-plus-delay ZCD circuit: input current shape (**a**,**b**) and its harmonic content (**c**,**d**) for converter in Table 1 at 115 Vac (**a**,**c**) and 230 Vac (**b**,**d**).

Compared to the previous case, the dead zone in $I_{AC}(\theta)$ is slightly narrower at high line: it occurs in the interval $-3.4° < \theta < 3.4°$ at 115 Vac and in $-6.2° < \theta < 6.2°$ at 230 Vac. To summarize:

- The optimal ZCD circuit does not alter the shape of the input current determined by the control mechanism but its implementation in a control IC requires a high level of silicon use.
- The differentiator-based ZCD circuit provides the highest THD values and its implementation in a control IC, though simple, requires silicon consuming structures able to withstand relatively large positive and negative voltages.
- The comparator-plus-delay ZCD circuit performs only slightly worse than the optimal ZCD circuit, provides the fine-tuning capability, and compared to the other two solutions, is less silicon consuming.

It is worth noticing that all the three ZCD circuit implementation do not alter the shape of the input current in the region $V_{IN} > V_R$; their difference in THD performance comes from the different behavior in the $V_{IN} \leq V_R$ region. A significant portion of this difference is in the amplitude of the dead-zone in $I_{AC}(\theta)$ that is generated. The diagram of Figure 20 shows how the total amplitude of the dead-zone changes as a function of the parameter $Kv = V_{PK}/V_R$ for various implementations of the ZCD circuit.

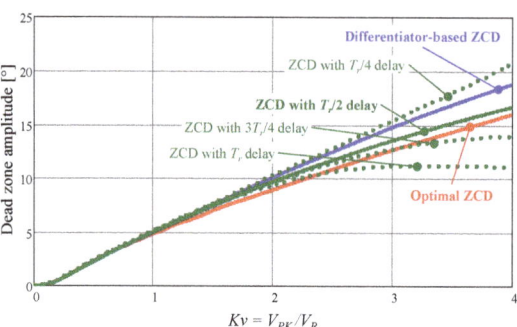

Figure 20. Dead-zone amplitude with various types of ZCD circuit as a function of the V_{PK}/V_R ratio.

It has been shown that the detrimental effect of the ZCD circuit on the input current shape is caused by a negative current in the turn-on instant of the power switch M that makes $T_{pos}(\theta) < T_{ON}(\theta)$. If we artificially delay the turn-on instant beyond $T_r/2$ after demagnetization, the initial current will be positive, thus making $T_{pos}(\theta) > T_{ON}(\theta)$. This will create a positive term (increasing with the extra delay) that will partly compensate for the negative term due to the ringing current and is expected to result in a lower THD of the input current. Actually, this is shown in the diagram of Figure 13 and is consistent with the reduction of the dead-zone amplitude and the resulting crossover distortion shown in Figure 20 with an increasing delay.

In conclusion, the comparator-plus-delay ZCD circuit seems to be the best practical choice, due to its fine-tuning capability, good performance, and simplicity. The comparator-plus-delay ZCD circuit might even outperform the optimal ZCD circuit when $T_{pos}(\theta) > T_{ON}(\theta)$. Anyway, any improvement in the THD attempted in this way should be traded off against the consequences of a long delay in restarting a new switching cycle: losing exact valley switching (higher turn-on losses and electromagnetic noise) and pushing the operation more deeply into DCM (worsening of current form factor, higher conduction losses).

5. Experimental Verifications

A pair of test benches have been set up to experimentally assess the impact of a nonzero current at turn-on on the THD of the input current to verify the theoretical predictions outlined in Section 4.

The first test bench was based on the reference Hi-PF QR flyback converter specified in Table 4 and whose picture is shown in Figure 21 on the left-hand side. The converter is an old design based on the L6562A, a PFC IC from STMicroelectronics primarily intended for boost-based PFC converters, that implements the QR control method.

The second test bench was based on the reference Hi-PF QR flyback converter specified in Table 5 and whose picture is shown in Figure 21 on the right-hand side. The converter is a newly developed design based on the HVLED007, a PFC IC from STMicroelectronics specific for flyback topology that implements the EQR control method.

The instrumentation used to set up the test benches included an ac source Chroma 61501 and an e-load Chroma 6314A + 63108A set in constant current mode (both converters provide a regulated output voltage); voltage waveforms acquisitions and time measurements were done with the oscilloscope Tektronix DPO 7054C; the THD was measured with a power meter Yokogawa WT210.

Figure 21. 35 W Hi-PF QR flyback controlled with QR method by the L6562A IC (**a**); 35 W Hi-PF QR flyback controlled with EQR method by the HVLED007 IC (**b**).

Both controllers have a comparator-plus-delay ZCD circuit onboard and in both cases, the delay from the transformer's demagnetization instant to the turn-on instant of the power switch M has been adjusted by acting on the external interface circuit between the auxiliary winding and the ZCD input pin, as shown in Figure 22.

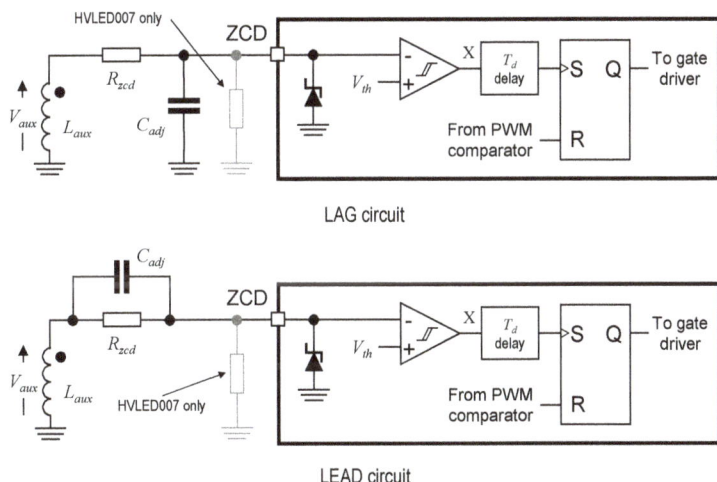

Figure 22. External circuits are used to adjust the delay between the transformer's demagnetization instant to turn-on instant of the power switch M. Upper circuit delays turn-on, lower circuit brings it forward.

It is worth reminding that the experimental data will provide the total result of all the concurrent causes of distortion and it is not generally possible to isolate each of them. Additionally, they are interacting with each other, thus a change in one of them may affect the amount of distortion caused by another one in a way that may be either detrimental or ameliorative. The objective is, therefore, to possibly capture a trend.

Experiments have been carried out at full load, where the contribution of other distortion causes (primarily, C_{in}) is expected to be at a minimum, at 115 Vac and 230 Vac.

Figure 23 shows a few key waveforms (drain-source voltage V_{ds}, auxiliary winding voltage V_{aux} and the gate-drive output V_{gs}) at the lower and upper ends of the adjustment range in the converter specified in Table 4 and controlled by the L6562A control IC with the QR method.

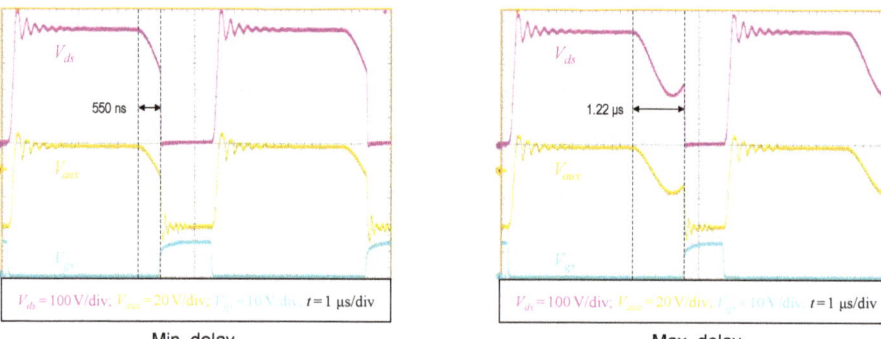

Figure 23. Key waveforms at the ends of the adjustment range of the turn-on instant t_{ON} for the converter specified in Table 4 and controlled by the L6562A control IC with the QR method.

The period of the ringing after demagnetization is 1.72 µs, then $T_r/2$ = 860 ns and the explored range (730 ns to 1.16 µs) includes both conditions $t_{ON} < T_{neg}$ and $t_{ON} > T_{neg}$ for $V_{IN} > V_R$.

Figure 24 shows (round markers connected by solid lines) the measured THD values of the input current $I_{AC}(\theta)$ as a function of different t_{ON} instants, obtained by varying C_{adj}, at both 115 Vac and 230 Vac and 100% load. For comparison, the plot shows also the corresponding values obtained by calculation (rhomboid markers connected by dotted lines) shown in the plot of Figure 12.

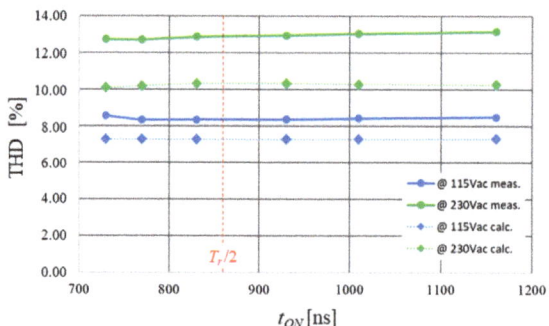

Figure 24. Plot of THD of the input current I_{AC} for different values of the delay between transformer's demagnetization instant to the turn-on instant t_{ON} of the power switch M for the converter specified in Table 4 and controlled by the L6562A control IC with the QR method. Theoretical values are shown for reference.

The experimental data confirm the "flat" trend of THD vs. t_{ON} predicted by the theoretical analysis, even though there is some discrepancy, a sort of offset, in the predicted values. The reasons for this difference have not been investigated but they might be explained by the presence of other distortion causes (e.g., C_{in}, or a negative input offset of the PWM comparator).

Figure 25 shows the same key waveforms (V_{ds}, V_{aux}, and V_{gs}) as in Figure 23 at the lower and upper ends of the adjustment range in the converter specified in Table 5 and controlled by the HVLED007 control IC with the EQR method.

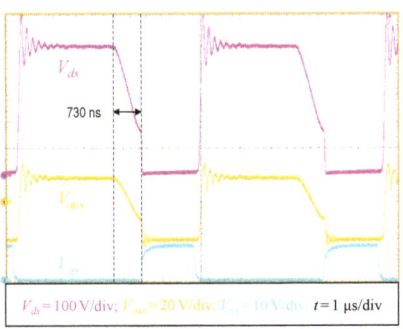
Min. delay

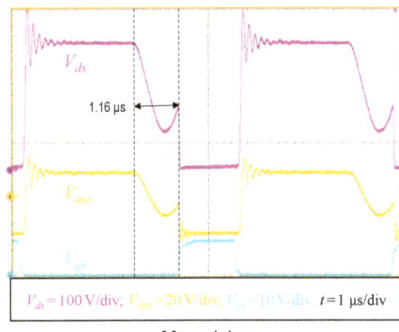
Max. delay

Figure 25. Key waveforms at the ends of the adjustment range of the turn-on instant t_{ON} for the converter specified in Table 5 and controlled by the HVLED007 control IC with the EQR method.

The period of the ringing after demagnetization is 2.07 µs, then $T_r/2 = 1.035$ µs and the explored range (550 ns to 1.22 µs) includes both conditions $t_{ON} < T_{neg}$ and $t_{ON} > T_{neg}$ for $V_{IN} > V_R$.

Figure 26 shows (round markers connected by solid lines) the measured THD values of the input current $I_{AC}(\theta)$ as a function of different t_{ON} instants, obtained by varying C_{adj}, at both 115 Vac and 230 Vac and 100% load. For comparison, the plot shows also the corresponding values obtained by calculation (rhomboid markers connected by dotted lines) reported in the plot of Figure 13.

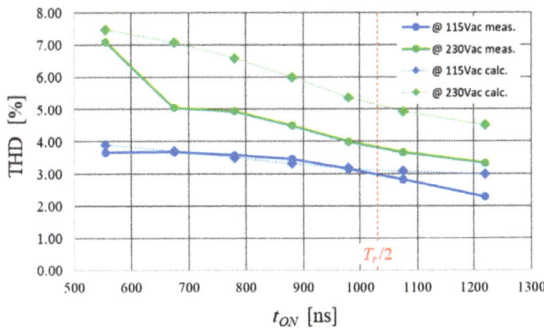

Figure 26. Plot of THD of the input current I_{AC} for different values of the delay between transformer's demagnetization instant to the turn-on instant t_{ON} of the power switch M for the converter specified in Table 5 and controlled by the HVLED007 control IC with the EQR method. Theoretical values are shown for reference.

The experimental data at low line are very well aligned to those calculated, except for the longest delay where the actual THD trend and the predicted one seem to diverge. At the high line, surprisingly, the measured values are lower than the calculated ones. However, the trend is the same except for the shortest delay, where the two values are much closer to one another. This difference is compatible with a positive offset of the PWM comparator, which increases the positive contribution of the per-cycle charge Q_{pos} and then, tends to reduce the THD.

To complete the experimental analysis, it is worth measuring the impact of t_{ON} on converter's efficiency (h = P_{out}/P_{in}). With valley switching ($t_{ON} = T_r/2$ when $V_{IN} > V_R$) the per-cycle energy lost at turn-on is at a minimum and increases as t_{ON} moves in either direction. However, with a shorter t_{ON} the switching frequency increases slightly (so do capacitive and switching losses) but operation gets closer to transition and the current form factor improves slightly (so conduction losses are a bit lower). With a longer t_{ON} there are the opposite changes in power losses. Additionally, a positive initial current causes a small amount of switching losses at turn-on, as if the converter worked in slight CCM.

Figure 27 shows the measured efficiency values of the converter specified in Table 4 and controlled by the L6562A control IC with the QR method as a function of different t_{ON} instants, the same as those considered in the plot of Figure 24, at 115 Vac and 230 Vac, 100% and 50% load.

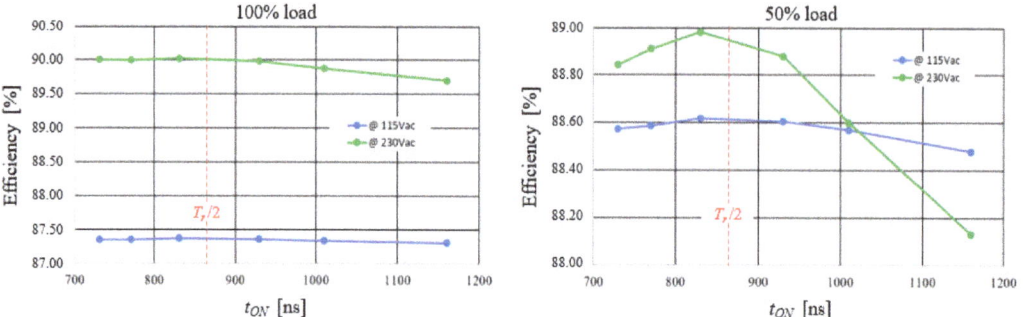

Figure 27. Plot of the measured efficiency for different values of the delay between transformer's demagnetization instant to the turn-on instant t_{ON} of the power switch M for the converter specified in Table 4 and controlled by the L6562A control IC with the QR method.

Figure 28 shows the measured efficiency values of the converter specified in Table 5 and controlled by the HVLED007 control IC with the EQR method as a function of different t_{ON} instants, the same as those considered in the plot of Figure 26, at 115 Vac and 230 Vac, 100% and 50% load.

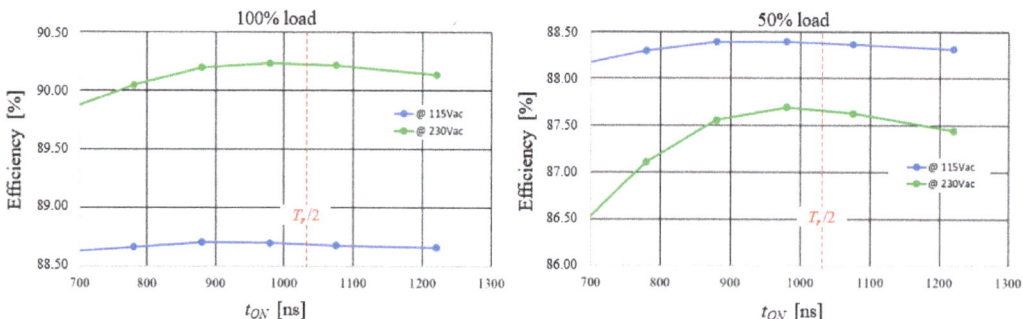

Figure 28. Plot of the measured efficiency for different values of the delay between transformer's demagnetization instant to the turn-on instant t_{ON} of the power switch M for the converter specified in Table 5 and controlled by the HVLED007 control IC with the EQR method.

In both cases it is possible to observe that the efficiency at low line is essentially insensitive to t_{ON}, which is a benign characteristic: since at low line the full load efficiency is at a minimum, i.e., power losses are at a maximum, the thermal design of the converter will be unaffected by the ZCD circuit, its setting, and its tolerances. This makes sense since under these conditions, conduction losses dominate.

At high line, where capacitive and switching losses dominate, the efficiency has a peak in the neighborhood of $t_{ON} = T_r/2$, meaning that capacitive losses are actually dominant. At 50% load this trend becomes more visible at low line as well.

These observations suggest a few system-level design guidelines. In converters controlled with the QR method, since t_{ON} has essentially no impact on the THD of the input current, it makes sense to set the ZCD circuit to target $t_{ON} = T_r/2$ to optimize their efficiency.

In converters controlled with the EQR method one can aim to minimize the THD of the input current by setting the ZCD circuit to target $t_{ON} > Tr/2$ with no impact on the thermal design. However, by doing so the drop in efficiency at high line and/or lighter load will be more pronounced. This fact should be kept in mind in designs where the electrical specification sets efficiency targets at high line and/or light load as well.

6. Conclusions

Hi-PF QR Flyback is the preferred converter different application thanks to the related high benefit/cost ratio. Although, one can implement an optimal control there are some inherent causes of distortion due to the nonideality of the components. In this work, the distortion due to the actual ZCD circuit has been qualitatively and quantitatively investigated. Moreover, a comparison among three ZCD has been performed. An optimal ZCD circuit does not negatively affect the input current but requires much silicon for implementing the control IC. A differentiator-based ZCD is the worst in terms of THD and, similarly to the previous case, it is silicon consuming. The comparator-plus-delay ZCD circuit enables the best trade-off between performance (only slightly worse than the optimal ZCD circuit) and silicon consumption (the lowest one). Two test benches have been used to experimentally assess the impact of a nonzero current at turn-on on the THD of the input current and verify the aforesaid theoretical predictions, finding a good agreement especially as far as the trend is concerned. The impact of a nonzero current at turn-on on efficiency has been assessed too.

As a conclusion, it is possible to state that the impact of the ZCD circuit on the THD of the input current in Hi-PF QR flyback converters is essentially negligible with the traditional QR method and low with the enhanced QR method. In other words, the detection method, as well as the quality and the performance of the semiconductor components utilized for the ZCD circuit, only slightly affect the THD of the input current or the efficiency of the converter. Therefore, utilizing more sophisticated and/or costly detection circuits does not necessarily provide significant improvement. Rather, it appears that the low-cost comparator-plus-delay ZCD circuit in use in the control ICs considered for the experiments is all in all the best choice.

The experiments have shown also that with the traditional QR method it is possible to set the comparator-plus-delay ZCD circuit to target the maximum conversion efficiency with no impact on the THD of the input current by setting $t_{ON} = Tr/2$. With the enhanced QR method it is possible to have a slight improvement of the THD by setting the ZCD circuit so as to have $t_{ON} > Tr/2$, with no penalty on the thermal design but with a higher deterioration rate of the efficiency at high line and/or light load.

Author Contributions: Writing-review and editing, C.A., G.G., S.A.R., G.S. All authors have read and agreed to the published version of the manuscript.

Funding: This research received no external funding.

Institutional Review Board Statement: Not applicable.

Informed Consent Statement: Not applicable.

Data Availability Statement: Not applicable.

Acknowledgments: The authors appreciate the help of Gianluca De Grandis from STMicroelectronics' Industrial and Power Conversion Division Application Laboratory for setting up the bench experiments.

Conflicts of Interest: The authors declare no conflict of interest.

References

1. Zhang, M.; Jovanovic, M.; Lee, F. Design considerations and performance evaluations of synchronous rectification in flyback converters. *IEEE Trans. Power Electron.* **1998**, *13*, 538–546. [CrossRef]
2. Elasser, A.; Torrey, D.A. Soft switching active snubbers for DC/DC converters. *IEEE Trans. Power Electron.* **1996**, *11*, 710–722. [CrossRef]
3. Kwon, J.-M.; Choi, W.-Y.; Kwon, B.-H. Single-Stage Quasi-Resonant Flyback Converter for a Cost-Effective PDP Sustain Power Module. *IEEE Trans. Ind. Electron.* **2011**, *58*, 2372–2377. [CrossRef]

4. Tsou, M.-C.; Kuo, M.-T. Optimal Combination Design of a Light Emitting Diode Matrix Applicable to a Single-Stage Flyback Driver. *Energies* **2020**, *13*, 5209. [CrossRef]
5. Yau, Y.-T.; Hwu, K.-I.; Liu, K.-J. AC–DC Flyback Dimmable LED Driver with Low-Frequency Current Ripple Reduced and Power Dissipation in BJT Linearly Proportional to LED Current. *Energies* **2020**, *13*, 4270. [CrossRef]
6. Nassary, M.; Orabi, M.; Arias, M.; Ahmed, E.M.; Hasaneen, E.-S.A. Analysis and Control of Electrolytic Capacitor-Less LED Driver Based on Harmonic Injection Technique. *Energies* **2018**, *11*, 3030. [CrossRef]
7. Adragna, C.; Gritti, G.; Raciti, A.; Rizzo, S.A.; Susinni, G. Analysis of the Input Current Distortion and Guidelines for Designing High Power Factor Quasi-Resonant Flyback LED Drivers. *Energies* **2020**, *13*, 2989. [CrossRef]
8. Mohamadi, A.; Afjei, E. A single-stage high power factor LED driver in continuous conduction mode. In Proceedings of the The 6th Power Electronics, Drive Systems & Technologies Conference (PEDSTC2015), Tehran, Iran, 3–4 February 2015; pp. 462–467.
9. Jeng, S.L.; Peng, M.T.; Hsu, C.Y.; Chieng, W.-H.; Shu, J.P.H. Quasi-Resonant Flyback DC/DC Converter Using GaN Power Transistors. *World Electr. Veh. J.* **2012**, *5*, 567–573. [CrossRef]
10. Vecchia, M.D.; Van den Broeck, G.; Ravyts, S.; Tant, J.; Driesen, J. Modified step-down series-capacitor buck converter with insertion of a Valley-Fill structure. *IET Power Electron.* **2019**, *12*, 3306–3314. [CrossRef]
11. Duan, R.-Y.; Lee, J. High-efficiency bidirectional DC-DC converter with coupled inductor. *IET Power Electron.* **2012**, *5*, 115. [CrossRef]
12. Cheng, C.-A.; Cheng, H.-L.; Chung, T.-Y. A novel single-stage high-power-factor LED street-lighting driver with coupled inductors. In Proceedings of the 2013 IEEE Industry Applications Society Annual Meeting, Lake Buena Vista, FL, USA, 6–11 October 2013; pp. 1–7.
13. Zhang, S.; Liu, X.; Guan, Y.; Yao, Y.; Alonso, J.M. Modified zero-voltage-switching single-stage LED driver based on Class E converter with constant frequency control method. *IET Power Electron.* **2018**, *11*, 2010–2018. [CrossRef]
14. Tabisz, W.; Gradzki, P.; Lee, F. Zero-voltage-switched quasi-resonant buck and flyback converters-experimental results at 10 MHz. *IEEE Trans. Power Electron.* **1989**, *4*, 194–204. [CrossRef]
15. Wong, K. *How to Choose Switching Controller for Design*; ON Semiconductor: Phoenix, AZ, USA, 2005; Appl. Note AND8205.
16. Sayani, M.; White, R.; Nason, D.; Taylor, W. Isolated feedback for off-line switching power supplies with primary-side control. In Proceedings of the APEC '88 Third Annual IEEE Applied Power Electronics Conference and Exposition, New Orleans, LA, USA, 1–5 February 1988; pp. 203–211. [CrossRef]
17. Imam, A.; Antony, B. Digitally controlled improved THD and power factor single-stage flyback LED driver with active input-current wave-shaping. In Proceedings of the 2013 Twenty-Eighth Annual IEEE Applied Power Electronics Conference and Exposition (APEC), Long Beach, CA, USA, 17–21 March 2013; pp. 3338–3344.
18. Shen, J.; Wu, Y.; Liu, T.; Zheng, Q. Constant current LED driver based on flyback structure with primary side control. In Proceedings of the 2011 IEEE Power Engineering and Automation Conference, Wuhan, China, 8–9 September 2011; pp. 260–263. [CrossRef]
19. Dong, H.J.; Xie, X.G.; Peng, K.S.; Li, J.S.; Zhao, C. A variable-frequency one-cycle control for BCM flyback converter to achieve unit power factor. In Proceedings of the IECON 2014—40th Annual Conference of the IEEE Industrial Electronics Society, Dallas, TX, USA, 29 October–1 November 2014; pp. 1161–1166.
20. Chern, T.-L.; Liu, L.-H.; Pan, P.-L.; Lee, Y.-J. Single-stage Flyback converter for constant current output LED driver with power factor correction. In Proceedings of the 4th IEEE Conference on Industrial Electronics and Applications, Xian, China, 25–27 May 2009; pp. 2891–2896.
21. Shagerdmootaab, A.; Moallem, M. Filter Capacitor Minimization in a Flyback LED Driver Considering Input Current Harmonics and Light Flicker Characteristics. *IEEE Trans. Power Electron.* **2014**, *30*, 4467–4476. [CrossRef]
22. Chou, H.-H.; Hwang, Y.-S.; Chen, J.Y. An Adaptive Output Current Estimation Circuit for a Primary-Side Controlled LED Driver. *IEEE Trans. Power Electron.* **2012**, *28*, 4811–4819. [CrossRef]
23. Kang, S.H.; Maksimovic, D.; Cohen, I. Efficiency Optimization in Digitally Controlled Flyback DC–DC Converters Over Wide Ranges of Operating Conditions. *IEEE Trans. Power Electron.* **2012**, *27*, 3734–3748. [CrossRef]
24. Jin, L.P.; Zhang, Y.C.; Jin, Y.Q. One stage flyback-type power factor correction converter for LED driver. In Proceedings of the International Conference on Electrical Machines and Systems, Busan, Korea, 26–29 October 2013; pp. 2173–2176.
25. Nie, W.; Zhu, W.; Ma, X.; Yu, Z.; Weidong, N.; Weimin, Z.; Xiaohui, M.; Zongguang, Y. A simple method to reduce line current zero-crossing distortion (LCZCD) for single-stage flyback LED driver. In Proceedings of the 12th IEEE International Conference on Solid-State and Integrated Circuit Technology (ICSICT), Guilin, China, 28–31 October 2014; pp. 1–3.
26. Mi, N.L.; Chung, R.; Jing, X.K. Design high power factor high efficiency primary-side regulated flyback LED driver. In Proceedings of the International Exhibition and Conference for Power Electronics, Intelligent Motion, Renewable Energy and Energy Management, Shanghai, China, 24–26 June 2015; pp. 218–225.
27. Adragna, C. *Design Equations of High-Power-Factor Flyback Converters based on the L6561*; STMicroelectronics: Grenoble, France, 2003; AN1059; Available online: https://www.st.com/resource/en/application_note/cd00004040-design-equations-of-highpowerfactor-flyback-converters-based-on-the-l6561-stmicroelectronics.pdf (accessed on 10 January 2021).
28. Adragna, C.; Gritti, G. High-power-factor quasi-resonant flyback converters draw sinusoidal input current. In Proceedings of the 2015 IEEE Applied Power Electronics Conference and Exposition (APEC), Charlotte, NC, USA, 15–19 March 2015; pp. 498–505.

Article

Analysis of the Usefulness Range of the Averaged Electrothermal Model of a Diode–Transistor Switch to Compute the Characteristics of the Boost Converter

Paweł Górecki * and Krzysztof Górecki

Department of Marine Electronics, Gdynia Maritime University, Morska 83, 81-225 Gdynia, Poland; k.gorecki@we.umg.edu.pl
* Correspondence: p.gorecki@we.umg.edu.pl

Abstract: In the design of modern power electronics converters, especially DC-DC converters, circuit-level computer simulations play an important role. This article analyses the accuracy of computations of the boost converter characteristics in the steady state using an electrothermal averaged model of a diode–transistor switch containing an Insulated Gate Bipolar Transistor (IGBT) and a rapid switching diode. This model has a form of a subcircuit for SPICE (Simulation Program with Integrated Circuit Emphasis). The influence of such factors as the switching frequency of the transistor, the duty cycle of the signal controlling the transistor, the input voltage, and the output current of the boost converter on the accuracy of computing the converter output voltage and junction temperature of the IGBT and the diode were analysed. The correctness of the computation results was verified experimentally. Based on the performed computations and measurements, the usefulness range of the model under consideration was determined, and a method of solving selected problems limiting the accuracy of computations of the characteristics of this converter was proposed.

Keywords: DC-DC converter; IGBT; averaged model; electrothermal model; SPICE; power electronics

1. Introduction

DC-DC converters [1–3] are commonly used to supply power electronics devices. One of the most popular types of these converters is a boost converter. This converter allows obtaining the output voltage higher than the input voltage while maintaining high conversion efficiency [1–3]. The diagram of the considered converter is shown in Figure 1.

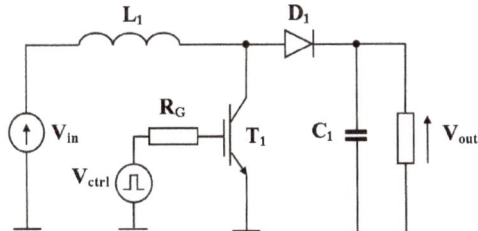

Figure 1. Diagram of a boost converter.

At the designing stage of all electronic systems, including DC-DC converters, computer simulations are necessary [1,4,5]. One of the most popular simulation programs used in electronics is SPICE (Simulation Program with Integrated Circuit Emphasis) [4,6]. One of its greatest advantages, compared to other popular simulation environments [7], is the ability to easily implement any compact model of an electronic component. Unfortunately, the models included in the SPICE program are isothermal, i.e., they do not take into account thermal phenomena. In these models, the junction temperature of the component is constant and does not depend on the power dissipated in it. In order to take into account thermal phenomena in computations, the SPICE program formulates models called electrothermal in the form of subcircuits for this program. The second significant disadvantage of the models built into the SPICE program is their limited accuracy, resulting from the simplified modelling of some important physical phenomena occurring in the component, which for the Insulated Gate Bipolar Transistor (IGBT) was presented, among others, in the paper [8].

The methods of estimating the internal temperatures of semiconductor devices using the model-order reduction techniques are described in the paper [9]. In turn, the methods of behavioural electrothermal modelling of power devices are described in the paper [10], whereas the methods of electrothermal simulations of switch-mode power converters are considered in the paper [11].

Performing the transient analysis of DC-DC converters in SPICE using the built-in models of electronic components is time-consuming. As it results from the paper [9], the time required to perform the computations to an electrically steady state is equal to even 24 h. From the engineering point of view, this is an unacceptably long time. Therefore, for many years, scientists have been conducting research on new methods of analysis that will allow reducing this time to an acceptable value [10–16]. These methods consist of modifying the classic method of transient analyses of electronic networks [10,11,13]. They use special algorithms to predict values of voltages and currents at the steady state.

Yet, another way is to use the method of a DC analysis with the averaged models of semiconductor devices or the whole electronic systems [12–14,16,17]. Using this method, it is possible to achieve a significant reduction in the computation time while maintaining the accuracy of computations of static characteristics.

One of such methods is a method using the averaged model of the diode–transistor switch [12,16,17], which is included in each single-inductor DC-DC converter, e.g., a boost converter, the diagram of which is shown in Figure 1. In such a model, the observation is used that at operation of the converter, the diode and the transistor contained in it conduct the current alternately, and for each of these devices associated with subcircuits, the equations describing the average values of voltages and currents of the converter in the steady state are formulated [3,17]. The following assumptions are made when formulating the equations constituting the average models used in the analysis of DC-DC converters:

(a) The formulated equations apply to the converter operation in the steady state,
(b) The equations take into account only DC components (average values) of voltages and currents,
(c) The switching times of semiconductor devices are negligibly short.

The equations of the model are obtained by equating to zero the dependences describing the average values of voltage on inductors and the average values of the capacitor currents occurring in the modelled network.

Averaged models of the diode–transistor switch have been described in the literature for many years, and the results of computations performed using such models are described for different DC-DC converters [3,12,17–26]. However, until recently, such models were formulated only for converters containing ideal switches [3,17–22] or Metal Oxide Semiconductor Field Effect Transistors (MOSFETs) [12,23–26], although IGBTs are not less popular in this application [27–30]. It is worth noting that when using models of ideal switches, it is impossible to compute the junction temperature of the diode and the transistor operating in the DC-DC converter. Meanwhile, the information about the values of these temperatures is important for the designer of the system and it allows verifying the correctness of the design. A diode–transistor switch with an IGBT requires a different approach than for MOSFET, e.g., for modelling the output characteristics, which results from their different shapes.

The main advantage of the averaged electrothermal model of a diode–transistor switch is short time of computations, which typically does not exceed 0.1 s. Nonetheless, this kind of model has some drawbacks. These drawbacks include neglecting dynamic effects in the thermal behaviour, impossibility of monitoring voltage on the transistor gate to observe a possible dielectric breakdown, and omitting the parasitic capacitances of semiconductor devices. Therefore, the averaged models may be useful to designers who need a fast and accurate estimation of the converter characteristics and verification that the semiconductor devices operate within their safe operating area.

The papers [31,32] propose a model of a diode–transistor switch for the analysis of DC-DC converters containing an IGBT. The model presented in [31] is a simplified model. In this model, a simplified method of describing the output characteristics of the transistor was used, and the influence of thermal phenomena is not taken into account. The main disadvantages of the model from [31] were removed in its next version, which are presented in [32]. However, this model does not take into account some important physical phenomena that may significantly affect the accuracy of computations using this model. The correctness of the cited model was verified only in a limited range of load current changes and at one value of the control signal frequency equal to 10 kHz.

This article analyses the influence of the accuracy of selected factors on the accuracy of the computations of the boost converter characteristics made with the use of the model from [32] under various operating conditions. Section 2 presents the model under consideration, Section 3 describes the measurement system used to verify the correctness of this model. Furthermore, in Section 4, the computed and measured characteristics of the boost converter in a wide range of changes in the load current and frequency of the control signal are presented. We also comment on the reasons for the discrepancies between the results of computations and measurements. A method of reducing these discrepancies is proposed, and their effectiveness is demonstrated in practice.

2. Considered Model

As it was mentioned above, in order to compute the DC characteristics of DC-DC converters, the averaged electrothermal models of a diode–transistor switch can be effectively used. The considered electrothermal model of the diode–transistor switch proposed in [32] is based on the known concept, which was previously presented for the MOSFET and the ideal lossless switch described, among others, in the papers [16,24]. This model allows determining the DC characteristics of any DC-DC converter containing a diode–transistor switch and junction temperature of the semiconductor devices contained therein. The diagram of this switch is shown in Figure 2.

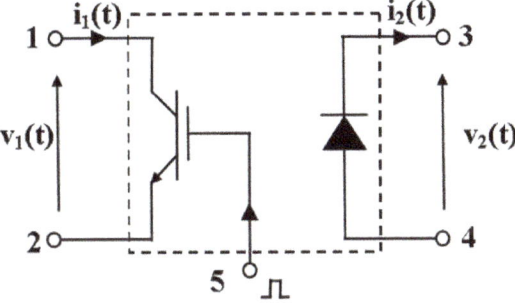

Figure 2. Diagram of a diode–transistor switch with the IGBT and the diode.

The network representation of the considered electrothermal averaged model of a diode–transistor switch (AVG model) is shown in Figure 3. This switch contains an IGBT and a diode.

Terminals 1, 2, 3, and 4 of the AVG model correspond to the terminals of the IGBT and the diode visible in the diagram of the diode–transistor switch shown in Figure 2. The considered model is connected to other components of the analysed DC-DC converters, using these terminals according to the respective diagrams. The frequency f and the duty cycle d of the signal controlling a DC-DC converter are parameters of the AVG model. These parameters that are used occur in the formulas describing some controlled voltage and current sources in this model. Internal temperatures of the IGBT and the diode correspond to voltages on terminals T_{jT} and T_{jD}, respectively.

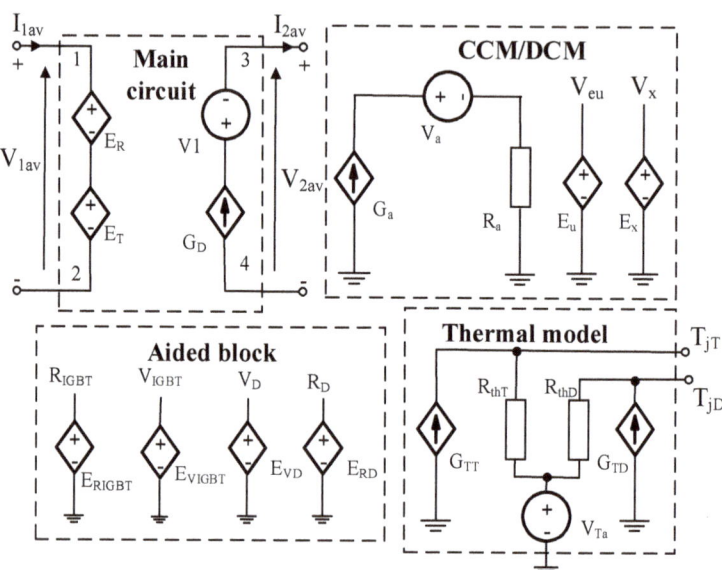

Figure 3. Network representation of the AVG model including the IGBT and the diode [32].

The described model contains four blocks: main circuit, aided block, thermal model, and CCM/DCM.

The main circuit contains two controlled voltage sources E_R and E_T and the controlled current source G_D. These voltage sources model the average voltage between the collector and the emitter of the IGBT, whereas the current source models the average value of the diode current. V1 is the voltage source of zero value that monitors this current. I_{1av} and I_{2av}, as well as V_{1av} and V_{2av} in the main circuit, denote the average values of voltages and currents of the semiconductor devices. The output values of the controlled voltage and current sources are described with the equations formulated using the idea presented in the papers [12,24]. According to this idea, at the steady state in each period T of the control signal, the current flows through one semiconductor device only. The transistor current flows for the time equal to d·T, and the diode current—for the time (1 − d)·T (in CCM (Continuous Conduction Mode) or less in DCM (Discontinuous Conduction Mode)). In the considered model, the current–voltage characteristics of both the semiconductor devices are described by piecewise linear functions. These functions use parameters depending on the internal temperatures of these semiconductor devices. Values of these parameters are computed in the aided block. In turn, internal temperatures of the diode and the IGBT are computed in the thermal model. The value of the equivalent duty cycle V_{eu} is computed in the controlled voltage source E_u included in the CCM/DCM block. The output value of this source depends on the parameters control signal—the duty cycle d and frequency f of the control signal, as well as inductance L of the inductor contained in the analysed DC-DC converter and load resistance of this converter.

The internal temperature of the IGBT (T_{jT}) and the diode (T_{jD}) are computed in the thermal model with self-heating phenomena taken into account. This model uses the electrical analogue of a DC compact thermal model [1,33]. In this analogue, voltages T_{jT}

and T_{jD} correspond to internal temperatures of the IGBT and the diode, respectively. In turn, the average values of power dissipated in these semiconductor devices are represented by the controlled current sources G_{TT} and G_{TD}. The output currents of these sources are described as follows [32].

$$G_{TD} = \left(V_D + \frac{R_D \cdot I_{2av}}{1 - V_{eu}}\right) \cdot \frac{I_{2av}}{1 - V_{eu}} \qquad (1)$$

$$G_{TT} = \left(V_{IGBT} + \frac{R_{IGBT} \cdot I_{1av}}{V_{eu}}\right) \cdot \frac{I_{1av}}{V_{eu}} \qquad (2)$$

where V_{IGBT}, V_D, R_D, and R_{IGBT} describe the parameters of piecewise linear models of the transistor and the diode characteristics.

The efficiency of the removal of heat generated in these devices is characterised by thermal resistance represented by resistors R_{thT} and R_{thD}. The voltage source V_{Ta} represents ambient temperature.

The other equations describing the controlled sources presented in Figure 3 are in [32].

3. Measurement Setup

In order to evaluate the usefulness of the considered model, the characteristics of the boost converter containing the considered semiconductor devices are measured and computed. The diagram of the investigated converter is presented in Figure 4.

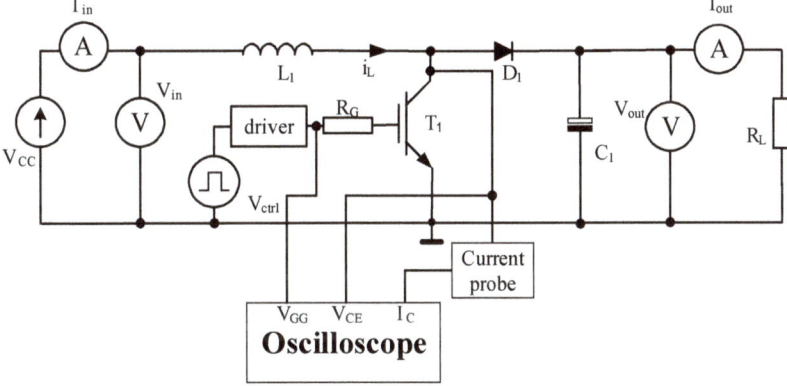

Figure 4. Diagram of the tested boost converter.

In the considered converter, the input voltage V_{CC} is supplied from the GWInstek PSB-2800L power supply. Appa 208 multimeters are used as ammeters and voltmeters. The inductance of coil L_1 is equal to 560 µH, and the capacitance of C_1 is 1 mF. The voltage controlling the IGBT V_{ctrl} is produced by the NDN C5603P function generator and connected to the gate of the IGBT via the driver IR2125. The prototype is mounted on the Printed Circuit Board (PCB). The diode of the type IDP08E65 is characterised by the maximum repetitive peak reverse voltage equal to 650 V and the maximum forward current equal to 16 A, and the IGBT of the type IGP06N60T is characterised by the maximum collector–emitter voltage equal to 600 V and the maximum DC collector current equal to 12 A. Based on the data given in the datasheet, the maximum switching frequency of this IGBT is estimated as equal to 500 kHz. Some waveforms of voltages and currents of the investigated DC-DC converter were measured using an oscilloscope Rigol MS05104 and a current probe Tektronix PCPA 300 for different values of load resistance and parameters of the control signal.

In order to determine the internal temperature T_j of the IGBT (T_{jT}) or the diode (T_{jD}), the case temperature T_C of these semiconductor devices is measured using a pyrometer

PT-3S by Optex. Next, using the values of thermal resistances junction-ambient R_{thj-a} and junction-case R_{thj-c}, the values of temperature T_j are calculated using the following formula [7]

$$T_j = T_c + (T_c - T_a) \cdot \frac{R_{thj-c}}{R_{thj-a} - R_{thj-c}} \qquad (3)$$

where T_a denotes ambient temperature. Values of thermal resistance R_{thj-c} are given in the data provided by the manufacturer, whereas thermal resistance R_{thj-a} is measured with the use of indirect electrical methods described e.g., in the paper [34].

The investigated DC-DC converter operates without any feedback loop, which is typically used in switch-mode power supplies including such converters. Therefore, at the selected operating conditions, the influence of changes in the load resistance and parameters of the control signal are not compensated for by the feedback loop. For this reason, any disadvantages of the considered model can be clearly illustrated for the investigated circuit.

4. Investigation Results

In order to analyse the usefulness range of the model from [32], hereinafter referred to as the AVG model, computations and measurements of the characteristics of the boost converter in various operating conditions are performed. In particular, the change of the following operating parameters of the considered converter is taken into account: input voltage V_{in}, load resistance R_L, frequency f, and the duty cycle d of the IGBT control signal. The measurements are carried out for a transistor and a diode operating at different cooling conditions. The operation of the investigated converter in both the CCM (Continuous Conduction Mode) and the DCM (Discontinuous Conduction Mode) [2] is considered.

The computations were performed for parameters with values identical to those given in [32], with the stipulation that the values of the model parameters R_{thD} and R_{thT} correspond to thermal resistances of the transistor and the diode. For the considered elements placed on heat sinks of the dimensions of 7.5 cm × 5 cm × 3.5 cm, the values of these parameters were 6.2 K/W, and for the devices operating without a heat sink, it was 43.85 K/W.

Selected results of measurements and computations of the investigated DC-DC converter are illustrated in the successive figures. In all these figures, points denote the results of measurements, whereas solid lines denote the results of computations performed with the AVG model.

Figure 5 shows the output characteristics of a boost converter operating with low input voltage V_{in} = 12 V under various cooling conditions, and Figure 6 shows the corresponding dependences of junction temperature of the transistor (a) and the diode (b) on the output current.

As it results from the data presented in Figures 5 and 6, the AVG model correctly reproduces the output characteristics of a boost converter operating in the CCM and the DCM, but only for the output current higher than 10 mA. For lower values of this current, there are significant discrepancies between the results of computations and measurements. The source of these discrepancies is losses resulting from switching off the IGBT omitted when formulating the considered model. The effect of these losses is an increase in the temperature of the transistor despite a decrease in the current value of the IGBT, as shown in Figure 6a. In order to improve the accuracy of modelling this characteristic, the heat generation model should take into account not only power losses related to the conduction of the IGBT but also power losses resulting from its switching. The method of modelling losses associated with the transistor switching for the needs of the averaged diode-transistor switch model is presented in [27] for the MOSFET transistor. In the range of very low values of the I_{out} current, the cause of the discrepancy may also be the omission of leakage currents of the transistor and the diode at the stage of formulating the considered model.

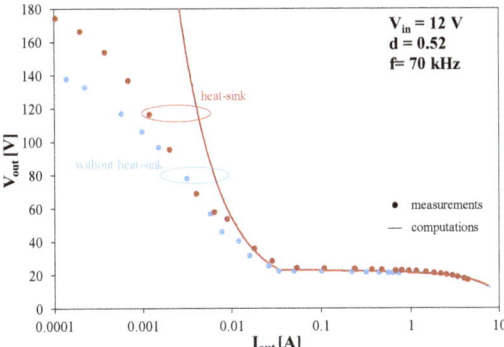

Figure 5. Measured and computed output characteristics of the boost converter for semiconductor devices operating at different cooling conditions.

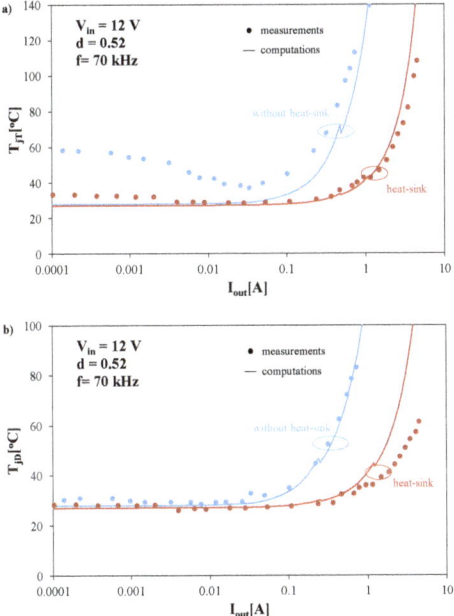

Figure 6. Measured and computed dependences of the IGBT junction temperature (**a**) and the diode junction temperature (**b**) on the output current of the investigated converter.

As shown in Figure 5, the cooling conditions do not significantly affect the shape of the $V_{out}(I_{out})$ dependence for $I_{out} > 1$ mA. However, they significantly affect the efficiency of heat removal from the semiconductor devices contained in the considered converter. As shown in the range of very low values of I_{out} current, an increase in temperature T_{jT} is visible, which proves that an average value of the dissipated power is as much as 0.5 W.

Figure 7 shows the output characteristics of the considered converter for four values of the IGBT switching frequency. Computations and measurements are performed for the transistor and the diode placed on heat sinks.

The results of computations and measurements presented in Figure 7 prove that for the collector currents below 10 mA and for switching frequencies higher than 10 kHz, the measurement results differ significantly from the results of computations made with the use of the AVG model. For the constant value of the output current, the discrepancy between the computations and measurements results increases with the switching frequency of the IGBT. An increase in the discrepancy also results from the fact that with an increase of switching frequency, the value of the current switched off by the transistor decreases [2]. A measure of this discrepancy is the relative error in determining voltage δV_{out} calculated from the formula

$$\delta V_{out} = \frac{|V_{outmeasured} - V_{outcomputed}|}{V_{outmeasured}} \quad (4)$$

where $V_{outmeasured}$ denotes the measured value of the output voltage, whereas $V_{outcomputed}$ is the value of this voltage computed with the use of the AVG model.

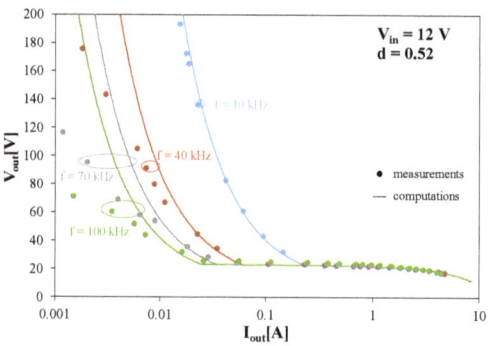

Figure 7. Measured and computed output characteristics of the boost converter for the transistor and the diode placed on heat sinks for selected values of switching frequency.

The computation error of the output voltage as a function of the output current is shown in Figure 8.

As it can be seen from the data presented in Figure 8, the relative error of determining the output voltage using the AVG model decreases with an increase of the output current of the boost converter and a decrease of frequency in the CCM. The error δV_{out} does not exceed a few percent. The graph has clear minima and maxima resulting from the modelling of the output characteristics of the turned-on IGBT using the piece-wise linear function. The smallest error values occur for a switching frequency equal to 10 kHz, and the highest error values occur for the highest among the considered frequencies, which is related to the AVG model not taking into account the influence of parasitic capacitances in the IGBT on the characteristics of the considered DC-DC converter.

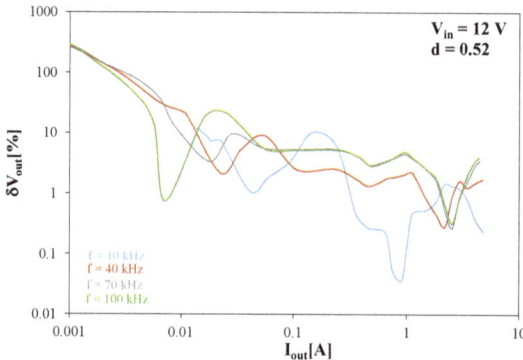

Figure 8. Computed values of the relative error in determining the output voltage as the function of the converter output current.

Figure 9 shows the dependence of the output voltage of the boost converter on the duty cycle of the signal controlling the transistor contained in it. During the measurements, the considered converter operated with the input voltage of 48 V. The IGBT and the diode were placed on the heat sinks, and the measurements were made for two values of load resistance. Additionally, in Figure 10a, the dashed line marks the results of computations made in the SPICE program using the AVG model, to which are added the values of the junction temperature increase ΔT_j resulting from dynamic losses in the IGBT computed on the basis of the Equations (5) and (6).

$$\Delta T_j = T_{jcond} + R_{thj-a} \cdot P_{swt} \qquad (5)$$

$$P_{swt} = \left(E_{on}\left(I_{on}, T_j, V_{CE}, R_G\right) + E_{off}\left(I_{off}, T_j, V_{CE}, R_G\right)\right) \cdot f \qquad (6)$$

where T_{jcond} is the value of the IGBT junction temperature computed using the AVG model, R_{thj-a} is the IGBT thermal resistance junction—ambient, and P_{swt} is the mean value of the power dissipated in the IGBT resulting from switching losses.

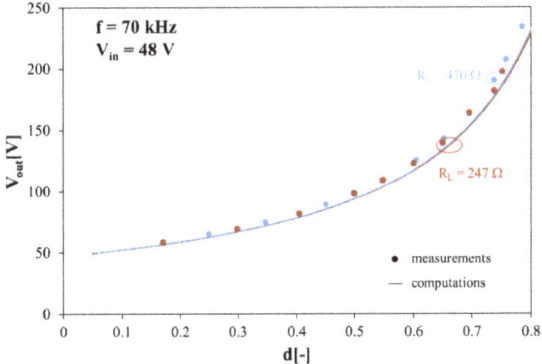

Figure 9. Measured and computed dependences of the output voltage on the duty cycle of the boost converter for selected values of load resistances.

The values of E_{on} and E_{off} necessary for the computations were obtained by approximating with linear functions the dependencies presented in the datasheet of the considered IGBT [35]. Due to the sawtooth shape of the collector current of the IGBT operating in the boost converter, it was necessary to separately determine the value of the switched currents I_{on} and I_{off}. The following formulas were used for this purpose [2].

$$I_{on} = I_{inavg} - \frac{\Delta I_L}{2} \qquad (7)$$

$$I_{off} = I_{inavg} + \frac{\Delta I_L}{2} \qquad (8)$$

$$\Delta I_L = \frac{(V_{out} - V_{in}) \cdot d}{L \cdot f}. \qquad (9)$$

The results presented in Figure 10a prove that even with the correctly determined values of the converter output voltage, which are shown in Figure 9, the computations of the corresponding values of the IGBT junction temperature may be characterised by a big error. The reason for this error is the omission of switching losses in the heat generation AVG model. In the case of the IGBT operation with high switching frequency, switching power losses that are proportional to e.g., to frequency and the switched current, can be much higher than the conduction losses of the IGBT [36,37]. This is demonstrated by the computation results presented in Figure 11, which show the conduction losses (red line) computed using the AVG model and the switching losses averaged for one period that are computed using the Formula (3)—blue line. The effect of frequency and the collector current on the average power dissipated in a transistor operating with resistive load is illustrated in Figure 11a,b, respectively.

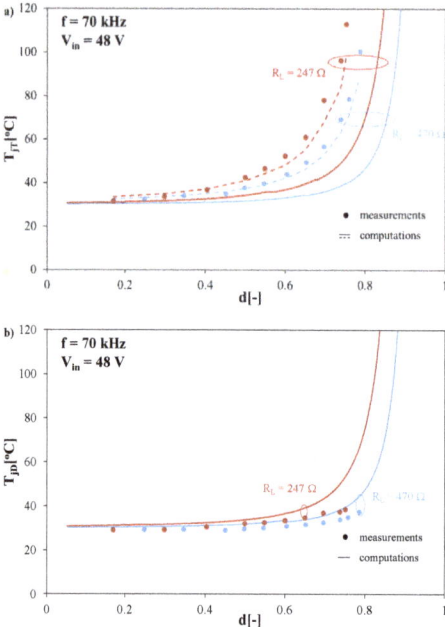

Figure 10. Measured and computed dependences of the IGBT junction temperature (**a**) and the diode junction temperature (**b**) on the duty cycle of the control signal of the IGBT in the investigated converter.

The curves shown in Figure 11 show that both switching and conduction losses are an important component of the power dissipated in this device during its operation. It is particularly noteworthy that in Figure 11 in the considered operating conditions, switching losses of the transistor are equal in the value to conduction losses for the frequency of 35 kHz, and for the frequency of 100 kHz, switching losses are already four times higher. It is also worth noting that in Figure 11b, for the considered operating conditions, switching losses depend more strongly on the collector current than on conduction losses.

The discrepancy between the computation results taking into account the impact of the IGBT switching losses on its junction temperature and the measurement results shown in Figure 11a probably results from not taking into account the influence of overvoltages and overcurrents on the value of turn-on and turn-off energies. Examples of overvoltages and overcurrents are shown in the waveforms shown in Figure 12.

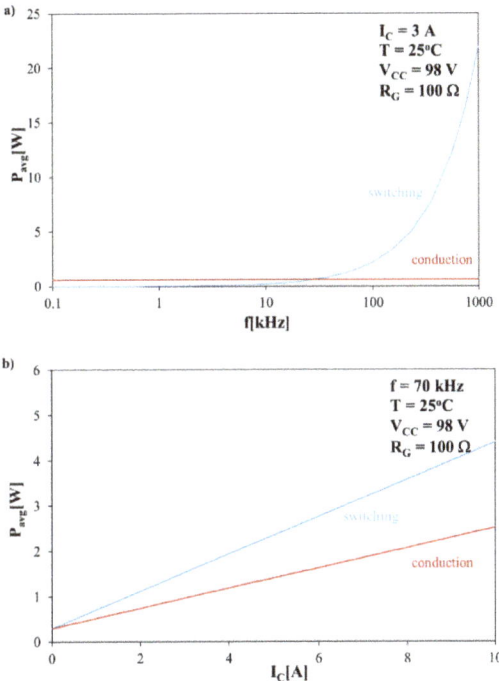

Figure 11. Computed dependences of the IGBT power losses caused by its switching (blue line) and conducting (red line) on frequency (**a**) and the collector current (**b**).

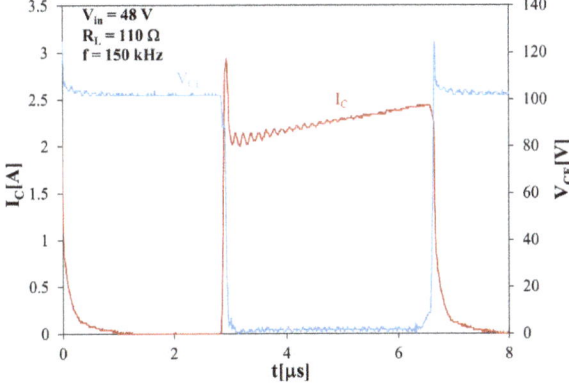

Figure 12. Measured waveforms of the collector–emitter voltage (blue solid line) and the collector current (red solid line) of the IGBT in the investigated boost converter.

In the case of a diode, in the considered frequency range, switching losses are much lower than conduction losses [38]. This is visible in Figure 10b, on the basis of which it can be stated that when using the diode model that does not take into account the effect of losses associated with its switching, the value of its junction temperature is computed with good accuracy.

Minor discrepancies between the results of computations and measurements, which are presented in Figure 9, result from inaccurate determination of the value of the duty cycle for the converter operating in the CCM. The problem of determining the value of the duty cycle is due to the fact that the operation of the IGBT with high switching frequency differs significantly from the operation of an ideal switch. In the case of an ideal switch, the value of the duty cycle voltage between the output terminals is equal to the value of the transistor control signal coefficient subtracted from one.

The computation method presented above was used in the formulas in the considered model. However, in the case of an IGBT operating with high switching frequency, due to non-zero values of parasitic capacitances, the value of the duty cycle of the signal between the output terminals of the transistor may significantly differ from the value for an ideal switch. In Figure 13, the dependences of the relative error in determining the duty cycle on the frequency of the control signal and the duty cycle of the control signal for the considered IGBT resulting from omitting the non-ideality of this device in determining this factor are presented. The value of this error was determined according to the formula

$$\delta d = \frac{|d_{GG} - (1 - d_{VCE})|}{1 - d_{VCE}} \qquad (10)$$

where d_{GG} is the value of the duty cycle of the signal at the output of the driver controlling the transistor, and d_{VCE} is the value of the duty cycle between the output terminals of the IGBT.

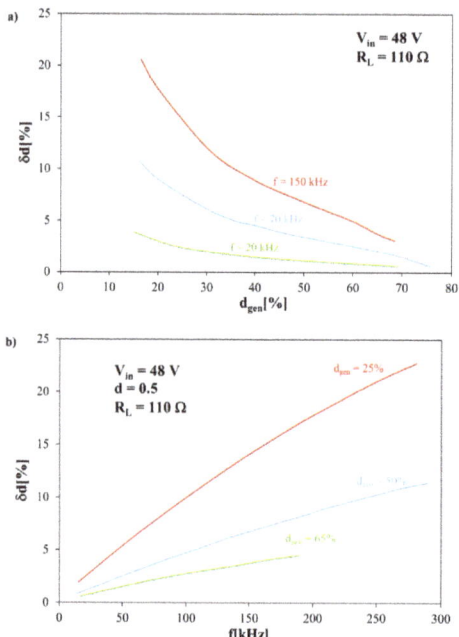

Figure 13. Measured values of the relative error in determining the duty cycle of the signal between the IGBT output terminals as a function of the duty cycle of the control signal (a) and the frequency of the signal controlling this transistor (b).

As it is presented in Figure 10, for low control signal duty cycle values, the user has poor control over the actual voltage duty cycle value between the transistor output terminals. A similar effect can be observed with increasing frequency. This can lead to significant discrepancies between the computed value of the converter output voltage from the mathematical model and the one obtained during the measurements of the real converter.

Figure 14 shows the dependence of the output voltage of the DC-DC converter on the switching frequency of the IGBT contained therein. In the figure, the computation results for the duty cycle value consistent with the signal at the driver output are marked using the solid line, and the dashed line is used for the V_{out} voltage values computed for the converter operating in the CCM taking into account the error in determining the duty cycle described above. In these computations, the dependence of the form is used:

$$V_{out} = V_{outc} + V_{in} \cdot \left(\frac{1}{1 - d_{GG} - \Delta d} - \frac{1}{1 - d_{GG}} \right) \quad (11)$$

where V_{outc} is the computed value of the output voltage using the model [33], and Δd is the absolute error in determining the duty cycle.

The converter output voltage dependencies presented in Figure 14 prove that neglecting the influence of electrical inertia related to the switching of the IGBT leads to a value of the duty cycle of the collector–emitter voltage different than for the ideal switch. This is the reason why the error in determining the output voltage increases with increasing frequency. At the same time, as the computations show, taking this into account in computations allows for a significant improvement of its accuracy.

As indicated in the papers [39,40], the omission of the nonlinear dependence of thermal resistance of the IGBT junction temperature significantly influences the accuracy of determining the transistor junction temperature. In order to check whether this problem also occurs for the transistor operating in the boost converter, the measurements and computations of its output characteristics and the correlation between the interior temperature of the IGBT operating without a heat sink were carried out. In order to emphasise the considered issue, Figure 15 presents only the measurement results for the converter operating in the CCM.

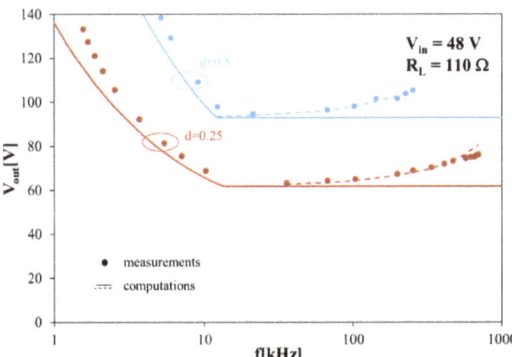

Figure 14. Measured and computed dependences of the output voltage on frequency.

The results presented in Figure 15 prove that even with the correct representation of electrical properties of the converter, i.e., losses related to switching the IGBT, the computed junction temperature value for high values of the output current is significantly overestimated due to the use of a linear thermal model in the AVG model. This problem could be solved relatively easily by averaging the nonlinear thermal model described in the paper [39]. In the modified model, the thermal capacitances will be omitted, and one nonlinear resistor describing the dependence $R_{thj-a}(T_j)$ will be used.

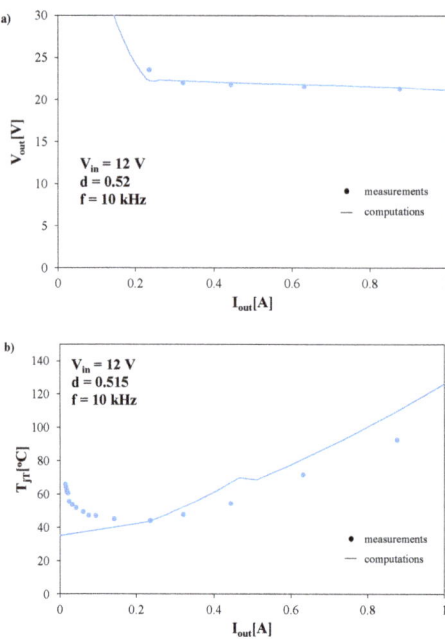

Figure 15. Measured and computed dependences of the output voltage (a) and the IGBT junction temperatures (b) on the output current of the investigated converter.

5. Conclusions

The paper presents the results of simulation and experimental studies illustrating the limitations of an electrothermal averaged model of a diode–transistor switch containing an IGBT and a rapid switching diode. The investigations were carried out for the boost converter in a wide range of changes of load current, frequency, and the duty cycle of the control signal. Various cooling conditions for the transistor and the diode contained in the tested converter were considered.

It was proved that for both the considered types of cooling conditions of semiconductor devices in the whole considered range of changes in the duty cycle and frequency of the control signal, good agreement was obtained between the computed and measured values of the output voltage of the boost converter operating in the CCM. On the other hand, in the DCM, the results of computations and measurements are convergent for the frequency $f < 20$ kHz and the load current $I_{out} > 10$ mA.

The problem of determining the value of the control signal duty cycle in the high-frequency range is pointed out. In this regard, the transistor switching times are not negligible in relation to the pulse duration. It was shown that at high frequency values $f > 70$ kHz, the error in determining the output voltage resulting from inaccuracy in determining the value of the duty cycle can reach even 10%. It is shown that this error could be reduced if the value of duty cycle is estimated, taking into account values of switch-on and switch-off times of the used transistor.

It is indicated that switching losses in the IGBT are significant for characteristics of the boost converter. It was shown that omitting these losses in the considered model causes significant, even twofold lowering of the value of the IGBT junction temperature increase above the ambient temperature. A modification of the description of the thermal model of the transistor using Formulas (5) and (6) was proposed to allow taking into account the influence of switching losses in the transistor on its junction temperature. It was shown that this modification allows improving the computation accuracy.

The problem of nonlinearity of thermal phenomena in semiconductor devices, which causes the junction temperature value to be overestimated in the range of high output currents of a converter, is also highlighted. The manner of taking into account the nonlinearity of the thermal model of the used semiconductor devices is proposed.

The problem of the junction temperature in the range of very low I_{out} values is also indicated. Improvement of the model under consideration that enables obtaining high accuracy of computations in a wide range of frequency changes will be the subject of further research by the authors.

The presented research results may be useful for designers of DC-DC converters. The presented results of computations and measurements also allow assessing the influence of the control signal frequency and load current as well as the cooling conditions of semiconductor devices on the properties of the tested boost converter.

Author Contributions: Conceptualisation, P.G. and K.G.; computations, P.G.; methodology, P.G. and K.G.; experimental verification, P.G.; writing—original draft preparation, P.G. and K.G.; writing—review and editing, K.G. and P.G.; visualisation, P.G.; supervision, K.G. All authors have read and agreed to the published version of the manuscript.

Funding: The scientific work is a result of the project No. 2018/31/N/ST7/01818 financed by the Polish National Science Centre.

Conflicts of Interest: The authors declare no conflict of interest.

References

1. Rashid, M.H. *Power Electronic Handbook*; Elsevier: Amsterdam, The Netherlands, 2007.
2. Kazimierczuk, M. *Pulse-Width Modulated DC-DC Power Converters*; Wiley: Hoboken, NJ, USA, 2015.
3. Ericson, R.; Maksimovic, D. *Fundamentals of Power Electronics, Norwell*; Kluwer Academic Publisher: Amsterdam, The Netherlands, 2001.
4. Rashid, M.H. *Spice for Power Electronics and Electric Power*; CRC Press: Boca Raton, FL, USA, 2006.
5. Maksimovic, D.; Stankovic, A.M.; Thottuvelil, V.J.; Verghese, G.C. Modeling and simulation of power electronics converters. *Proc. IEEE* **2001**, *89*, 898–912. [CrossRef]
6. Vladirmirescu, A. Shaping the History of SPICE. *IEEE Solid State Circuits Mag.* **2011**, *3*, 36–39.
7. Górecki, P.; Wojciechowski, D. Accurate Computation of IGBT Junction Temperature in PLECS. *IEEE Trans. Electron Devices* **2020**, *67*, 2865–2871. [CrossRef]
8. Górecki, P.; Górecki, K.; Zarębski, J. Modelling the temperature influence on dc characteristics of the IGBT. *Microelectron. Reliab.* **2017**, *79*, 96–103. [CrossRef]
9. Codecasa, L.; Catalano, A.P.; d'Alessandro, V. A priori Error Bound for Moment Matching Approximants of Thermal Models. *IEEE Trans. Compon.* **2019**, *9*, 2383–2392. [CrossRef]
10. D'Alessandro, V.; Codecasa, L.; Catalano, A.P.; Scognamillo, C. Circuit-Based Electrothermal Simulation of Multicellular SiC power MOSFETs using FANTASTIC. *Energies* **2020**, *13*, 4563. [CrossRef]
11. Catalano, A.P.; Riccio, M.; Codecasa, L.; Magnani, A.; Romano, G.; d'Alessandro, V.; Maresca, L.; Rinaldi, N.; Breglio, G.; Irace, A. Model-Order Reduction Procedure for Fast Dynamic Electrothermal Simulation of Power Converters. Applications in Electronics Pervading Industry, Environment and Society. *Lect. Notes Electr. Eng.* **2017**, *512*, 81–87.
12. Górecki, K.; Detka, K. Application of Average Electrothermal Models in the SPICE-Aided Analysis of Boost Converters. *IEEE Trans. Ind. Electron.* **2019**, *66*, 2746–2755. [CrossRef]
13. Górecki, K.; Zarębski, J. The Method of a Fast Electrothermal Transient Analysis of Single-Inductance DC-DC Converters. *IEEE Trans. Power Electron.* **2012**, *27*, 4005–4012. [CrossRef]
14. Krein, P.T.; Bentsman, J.; Bass, R.M.; Lesieutre, B.L. On the Use of Averaging for the Analysis of Power Electronic Systems. *IEEE Trans. Power Electron.* **1990**, *5*, 182–190. [CrossRef]
15. Vorperian, V. *Fast Analitycal Techniques for Electrical and Electronic Circuits*; Cambridge University Press: Cambridge, UK, 2002.
16. Ben-Yaakov, S.; Gaaton, Z. Generic SPICE compatible model of current feedback in switch mode converters. *Electron. Lett.* **1992**, *28*, 1356–1358. [CrossRef]
17. Basso, C. *Switch-Mode Power Supply SPICE Cookbook*; McGraw-Hill: New York, NY, USA, 2001.
18. Azer, P.; Rodriguez, R.; Guo, J.; Gareau, J.; Bauman, J.; Ge, H.; Bilgin, B.; Emadi, A. Time-Efficient Integrated Electrothermal Model for a 60-kW Three-Phase Bidirectional Synchronous DC-DC Converter. *IEEE Trans. Ind. Appl.* **2020**, *56*, 654–668. [CrossRef]
19. Han, J.; Zhang, B.; Qiu, D. Bi-switching Status Modeling Method for DC-DC Converters in CCM and DCM Operations. *IEEE Trans. Power Electron.* **2017**, *21*, 2464–2472. [CrossRef]
20. Pavlovic, T.; Bjazic, T.; Ban, Z. Simplified Averaged Models of DC-DC Power Converters Suitable for Controller Design and Microgrid Simulation. *IEEE Trans. Power Electron.* **2013**, *28*, 3266–3275. [CrossRef]

21. Mahdavi, J.; Emaadi, A.; Bellar, M.D.; Ehsani, M. Analysis of Power Electronic Converters Using the Generalized State-Space Averaging Approach. *IEEE Trans. Circuits Syst.* **1997**, *44*, 767–770. [CrossRef]
22. Azer, P.; Emadi, A. Generalized State Space Average Model for Multi-Phase Interleaved Buck, Boost and Buck-Boost DC-DC Converters: Transient, Steady-State and Switching Dynamics. *IEEE Access* **2020**, *8*, 77736–77745. [CrossRef]
23. Cheng, T.; Dah-ChuanLu, D.; Siwakoti, Y.P. Electro-Thermal Modeling of a Boost Converter Considering Device Self-heating. In Proceedings of the 2020 IEEE Applied Power Electronics Conference and Exposition (APEC), New Orleans, LA, USA, 15–19 March 2020.
24. Górecki, K. A new electrothermal average model of the diode-transistor switch. *Microelectron. Reliab.* **2008**, *48*, 51–58. [CrossRef]
25. Bryant, B.; Kazimierczuk, M.K. Voltage-Loop Power-Stage Transfer Functions with MOSFET Delay for Boost PWM Converter Operating in CCM. *IEEE Trans. Ind. Electron.* **2007**, *54*, 347–353. [CrossRef]
26. Ayachit, A.; Kazimierczuk, M.K. Averaged Small-Signal Model of PWM DC-DC Converters in CCM Including Switching Power Loss. *IEEE Trans. Circuits Syst.* **2019**, *66*, 262–266. [CrossRef]
27. Dimitrov, B.; Hayatleh, K.; Barker, S.; Collier, G.; Sharkh, S.; Cruden, A. A Buck-Boost Transformerless DC-DC Converter Based on IGBT Modules for Fast Charge of Electric Vehicles. *Electronics* **2020**, *9*, 397. [CrossRef]
28. Loncarski, J.; Monopoli, V.G.; Cascella, G.L.; Cupertino, F. SiC-MOSFET and Si-IGBT-Based dc-dc Interleaved Converters for EV Chargers: Approach for Efficiency Comparison with Minimum Switching Losses Based on Complete Parasitic Modeling. *Energies* **2020**, *13*, 4585. [CrossRef]
29. Zhao, Z.; He, X. Research on Digital Synchronous Rectification for a High-Efficiency DC-DC Converter in an Auxiliary Power Supply System of Magnetic Levitation. *Energies* **2020**, *13*, 51. [CrossRef]
30. Concari, L.; Barater, D.; Toscani, A.; Concari, C.; Franceschini, G.; Buticchi, G.; Liserre, M.; Zhang, H. Assessment of Efficiency and Reliability of Wide Band-Gap Based H8 Inverter in Electric Vehicle Applications. *Energies* **2019**, *12*, 1922. [CrossRef]
31. Górecki, P. Application of the Averaged Model of the Diode-transistor Switch for Modelling Characteristics of a Boost Converter with an IGBT. *Int. J. Electron. Telecommun.* **2020**, *66*, 555–560.
32. Górecki, P.; Górecki, K. Electrothermal Averaged Model of a Diode-Transistor Switch Including IGBT and a Rapid Switching Diode. *Energies* **2020**, *13*, 3033. [CrossRef]
33. Górecki, K.; Zarębski, J.; Górecki, P.; Ptak, P. Compact Thermal Models of Semiconductor Devices-a Review. *Int. J. Electron. Telecommun.* **2019**, *65*, 151–158.
34. Górecki, K.; Górecki, P.; Zarębski, J. Measurements of parameters of the thermal model of the IGBT module. *IEEE Trans. Instrum. Meas.* **2019**, *68*, 4864–4875. [CrossRef]
35. Low Loss IGBT: IGBT in TRENCHSTOP™ and Fieldstop Technology. Available online: https://www.infineon.com/dgdl/Infineon-IGP06N60T-DS-v02_03-en.pdf?fileId=db3a30432313ff5e0123b82d13ba7883 (accessed on 12 November 2020).
36. Blaabjerg, F.; Jaeger, U.; Munk-Nielsen, S.; Pedersen, J.K. Power Losses in PWM-VSI Inverter Using NPT or PT IGBT Devices. *IEEE Trans. Power Electron.* **1995**, *10*, 358–367. [CrossRef]
37. Li, Z.; Wang, J.; Ji, B.; Shen, Z.J. Power Loss Model and Device Sizing Optimization of Si/SiC Hybrid Switches. *IEEE Trans. Power Electron.* **2020**, *35*, 8512–8523. [CrossRef]
38. Starzak, L.; Stefanskyi, A.; Zubert, M.; Napieralski, A. Improvement of an electro-thermal model of SiC MPS diodes. *IET Power Electron.* **2018**, *11*, 660–667. [CrossRef]
39. Górecki, K.; Górecki, P. Nonlinear Compact Thermal Model of the IGBT Dedicated to SPICE. *IEEE Trans. Power Electron.* **2020**, *35*, 13420–13428. [CrossRef]
40. Codecasa, L.; d'Alessandro, V.; Magnani, A.; Irace, A. Circuit-Based Electrothermal Simulation of Power Devices by an Ultrafast Nonlinear MOR Approach. *IEEE Trans. Power Electron.* **2016**, *31*, 5906–5916. [CrossRef]

Article

A Universal Mathematical Model of Modular Multilevel Converter with Half-Bridge

Ming Liu [1,2], Zetao Li [1,3,*] and Xiaoliu Yang [1]

1. College of Electrical Engineering, Guizhou University, Guiyang 550025, China; weiminxiaohai@163.com (M.L.); xlyang1@gzu.edu.cn (X.Y.)
2. Department of Mechanical Engineering, Guizhou College of Electronic Science and Technology, Guian 550003, China
3. Guizhou Provincial Key Laboratory of Internet + Intelligent Manufacturing, Guiyang 550025, China
* Correspondence: gzulzt@163.com

Received: 30 June 2020; Accepted: 27 August 2020; Published: 29 August 2020

Abstract: Modular multilevel converters (MMCs) play an important role in the power electronics industry due to their many advantages, such as modularity and reliability. In the current research, the simulation method is used to study the system. However, with the increasing number of sub-modules (SMs), it is difficult to model and simulate the system. In order to overcome these difficulties, this paper presents a universal mathematical model (UMM) of MMC using half-bridge cells as SMs. The UMM is a full-scale model with switching state, capacitance, inductance, and resistance characteristics. This method can calculate any number of SMs, and it does not need to build a simulation model (SIM) of physical MMC—in particular, parametric design can be realized. Compared with the SIM, the accuracy of the proposed UMM is verified, and the computational efficiency of the UMM is 8.7 times higher than the simulation method. Finally, by utilizing the proposed UMM method, the influence of the parameters of MMCs is studied, including the arm induction, SM capacitance, SM number, and output current/voltage total harmonic distortion (THD) based on the UMM in the paper. The results offer an engineering insight to optimize the design of MMCs.

Keywords: modular multilevel converter (MMC); total harmonic distortion (THD); universal mathematical model (UMM); switching state; nearest level modulation (NLM)

1. Introduction

With the rapid development of offshore wind farms, the demand for a high power, high-quality transmission system becomes more urgent. Modular multilevel converter (MMC)-based high voltage direct voltage (HVDC) technology provides a promising solution, due to its advantages of modularization, scalability, high efficiency, excellent harmonic performance, fault blocking ability, small filter size, high efficiency, and low redundancy cost [1–3]. MMC has been applied to many industries, such as energy storage systems, medium-voltage and high-power motor drive systems, distribution systems, etc. [4–8].

However, it is difficult to formulate an explicit expression of MMC, because it is a hybrid system of discrete and continuous models. The main feature of MMCs is the cascaded connection of a large number of sub-modules (SMs). These SMs are arranged in groups called arms or branches. The low-frequency voltage or current at the AC side is controlled by high-frequency switching values to manage SMs on/off. Therefore, the interaction between the arm and line quantities (variables) generates low- and high-frequency components on the AC and DC side of the SMs in an MMC [9]. In other words, MMC has strong coupling nonlinear multi-input and multi-output dynamic features [10]. The simulation studies were utilized to analyze the behavior of an MMC. However, the simulation process consumes time and computer resources to create a large number of SMs (up to 400 per arm) [11].

For example, a traditional detailed model (TDM) of MMC requires hundreds and thousands of Insulated Gate Bipolar Transistors(IGBTs) with antiparallel diodes and capacitors to be built and electrically connected in the simulation package's graphical user interface, resulting in a large admittance matrix.

To simplify the simulation model, the conventional switching models/detailed models with full capabilities of replicating the conduction of power electronic devices such as IGBTs and their anti-parallel diodes are inefficient for the modeling of MMC-HVDC, as the simulation time is prohibitively long [1]. To simplify the calculation, it was assumed that the SM capacitor voltages are well balanced at their reference values [12,13]. The SM terminal voltage in each arm was modeled as a single equivalent voltage source [14]. In [15], the equivalent model was used and a small-signal analysis was carried out. Alternatively, each arm of MMC was modeled as a nonlinear capacitor with a time-variant sinusoidal capacitance [16]. Moreover, to simplify the analysis, the average value models (AVMs) are presented in [17–20]. However, the methods above do not reflect the switching state and the transient process of the SM capacitor voltage.

To address this problem, an efficient model was proposed by Udana and Gole in [21], which is referred to as the detailed equivalent model (DEM) in this paper; yet, a drawback of the DEM is that the individual converter components are invisible to the user. A new model, referred to as the accelerated model (AM), was proposed by Xu et al. in [22], but a full and objective comparison could not be completed because different researchers built the models on different computers. In [23], an enhanced accelerated model (EAM) with improved simulation speed was proposed by Antony et al., which further improved the computational efficiency of one method. A new dynamic phasor (DP) model of an MMC with an extended frequency range for direct interfacing with an electromagnetic transient (EMT) simulator was presented in [24]. In reference [25], a method of MMC modeling and design based on parametric and model-form uncertainty quantification is proposed, which can establish confidence in modeling and simulation in the presence of manufacturing variability and modeling errors, and may eliminate the need for heuristic safety factors. However, the high-efficiency calculation of large-scale SMs is not involved. The internal dynamics of the MMC are modeled considering the dominant harmonic components of each variable. However, the improved simulation models introduced above are based on Power Systems Computer Aided Design/Electromagnetic Transients including DC(PSCAD/EMTDC) for electromagnetic transient simulation. It is inconvenient to set variable parameters or change the topology of the whole circuit. It cannot be satisfied by loop calculation to compare the changes of parameters.

In this paper, a universal mathematical model for MMC is proposed which can reflect the steady-state and dynamic process of MMC. The model is a detailed numerical model, including the capacitor voltage and switch function of SMs. It can be implemented by a computer for any number of SMs. The whole model is parametric programmed; by setting one or several parameters, the desired results can be quickly obtained. Compared with the simulation model, this algorithm can easily modify the circuit parameters by setting cycle statements, and automatically carry out repeated simulation and multi-state simulation, so it is convenient to observe the operation characteristics of the system under different parameter values. Using the proposed MMC, the output voltage and the current THDs of MMC have been analyzed under different parameters (such as module number, capacitor voltage, arm inductance). In addition, the change in the capacitance voltage has been studied as one capacitance value decayed.

Without losing generality, in this paper the research is based on MATLAB/Simulink because of its powerful numerical calculation ability and rich processing module (such as Pulse Width Modulation (PWM) generator, and various transformation and comparison modules), especially the demonstration of control strategy. Compared with Simulink, PLECS is more professional, but does not have as many toolboxes as Simulink. The LTSpice installation package is small, easy to operate, and fast, but most of the support is for the ADI company's own chip model. PSIM has the advantages of simple operation and fast simulation speed, and supports mainstream simulation mode analysis. However, due to the use of ideal switches, the simulation accuracy is limited.

This paper is organized as follows. Section 2 introduces the MMC topology, operation, and mathematical model. The algorithm of the universal mathematical model (UMM) is explained in Section 3. The correctness of the UMM is demonstrated in Section 4. A performance analysis under different conditions is shown in Section 5. The conclusions of the study are presented in Section 4.

2. Topology and Mathematical Model of the MMC

Topology and Principles of Operation

The circuit structure of a three-phase MMC is shown in Figure 1. The single-phase consists of an upper and a lower arm. Each arm is composed of N sub-modules (SM), and an inductor and equivalent resistance are connected in series. Each individual SM contains a capacitor and two complementary insulated gate bipolar transistor modules (i.e., $S_{jm,n}$ and $S'_{jm,n}$). In this paper, the subscript $j = a, b, c$ means three-phase; $m = u, l$, where u represents the upper arm and l represents a lower arm; $n = 1, 2, 3, \ldots, N$ represents the number of sub-modules. The rest of the symbols are as follows: L (the arm inductance), R (the arm equivalent resistance), u_{jm} (the arm voltage), i_{jm} (the arm current), u_j (the AC side voltage), i_{oj} (the output current), i_{cj} (the circulating current), L_{oj} (the load inductance), R_{oj} (the load resistance), U_{dc} (the DC source voltage), $u_{jm,n}$ (the capacitor voltage).

$$\frac{du_{jm,n}}{dt} = \frac{S_{jm,n} i_{jm}}{C} \tag{1}$$

where $S_{jm,n}$ is the switch function of the nth SM in the m arm of phase-j, and its value is 1 or 0; C is the SM capacitance.

The relationship between the arm voltage and the capacitor voltage in phase-j and the switching function is:

$$u_{jm} = \sum_{n=1}^{N} S_{jm,n} u_{jm,n} \tag{2}$$

Considering a fictitious midpoint in the DC side of Figure 1 and using Kirchhoff's circuit laws, the following mathematical equations that govern the dynamic behavior of the MMC in phase-j can be obtained:

$$u_{ju} + Ri_{ju} + L\frac{di_{ju}}{dt} + R_{oj}i_{oj} + L_{oj}\frac{di_{oj}}{dt} + u_j - \frac{U_{dc}}{2} = 0 \tag{3}$$

$$u_{jl} + Ri_{jl} + L\frac{di_{jl}}{dt} - R_{oj}i_{oj} - L_{oj}\frac{di_{oj}}{dt} - u_j - \frac{U_{dc}}{2} = 0 \tag{4}$$

$$i_{oj} = i_{ju} - i_{jl} \tag{5}$$

The arm currents can be expressed as:

$$i_{ju} = \frac{i_{oj}}{2} + i_{cj}, \quad i_{jl} = -\frac{i_{oj}}{2} + i_{cj} \tag{6}$$

where i_{cj} is the circulating currents flowing through phase-j of the MMC and can be calculated by Equation (7):

$$i_{cj} = \frac{i_{ju} + i_{jl}}{2} \tag{7}$$

Substitute (3) and (5) into (4), and the dynamics of phase-j AC-side currents can be obtained as:

$$\frac{di_{oj}}{dt} = -\frac{R + 2R_{oj}}{L + 2L_{oj}}i_{oj} - \frac{1}{L + 2L_{oj}}u_{ju} + \frac{1}{L + 2L_{oj}}u_{jl} - \frac{2}{L + 2L_{oj}}u_j \tag{8}$$

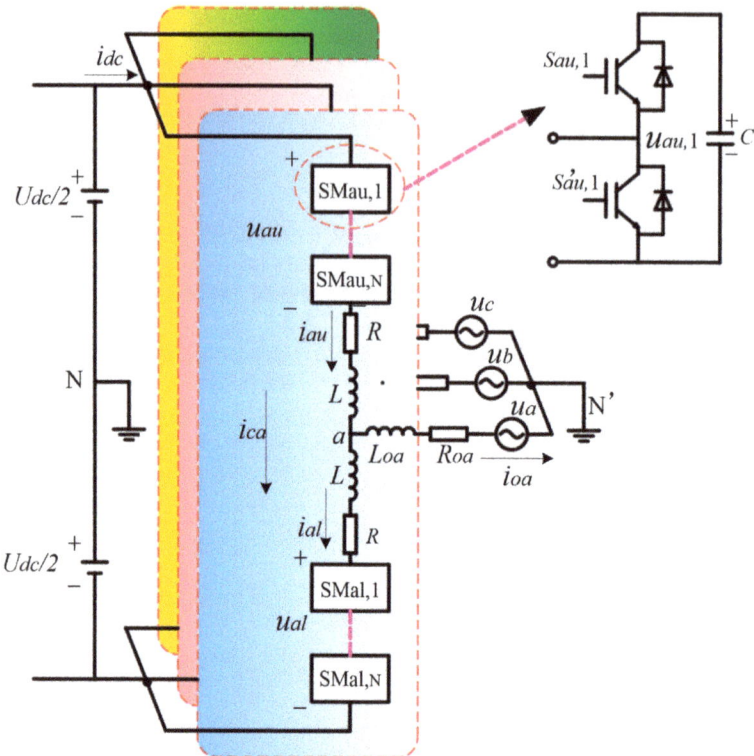

Figure 1. Structure of a three-phase MMC-based inverter and its SM.

Similarly, the dynamic behavior of the circulating current in phase-j can be obtained by substituting (3) and (7) into (4):

$$\frac{di_{cj}}{dt} = -\frac{R}{L}i_{cj} - \frac{1}{2L}u_{ju} - \frac{1}{2L}u_{jl} + \frac{1}{2L}U_{dc} \tag{9}$$

Based on (1), (2), (8), and (9), the state-space equation of the MMC in phase-j can be described as:

$$\dot{x}(t) = A(t)x(t) + Dd(t) \tag{10}$$

where $x = [i_{oj}, i_{cj}, u_{ju,1}, \ldots, u_{ju,N}, u_{jl,1}, \ldots, u_{jl,N}]^T \in \mathbb{R}^{2N+2}$ is the state vector; $d = [u_j, U_{dc}]^T$ is a perturbation vector; the desired value (u_j^*) of u_j is presented in Equation (11); $A \in \mathbb{R}^{(2N+2)\times(2N+2)}$ is a time-varying structure state matrix, presented in (12); and $D \in \mathbb{R}^{(2N+2)\times 2}$ is the perturbation coefficient matrix, presented in (20).

$$u_j^* = \sqrt{2}U\sin(2\pi f t + \varphi_j) \tag{11}$$

where U is the voltage effective value (RMS) on the AC side, f is the AC system frequency, and φ_j is the initial phase angle in phase-j.

$$A(t) = \begin{bmatrix} A_1 & A_2(t) \\ A_3(t) & 0 \end{bmatrix} \in \mathbb{R}^{(2N+2)\times(2N+2)} \tag{12}$$

where $A_1 \in \mathbb{R}^{2\times2}$ is a constant matrix in (13), and $A_2 \in \mathbb{R}^{2\times2N}$ and $A_3 \in \mathbb{R}^{2N\times2}$ are time-varying matrixes in (14) and (18), respectively.

$$A_1 = \begin{bmatrix} -\frac{R + 2R_{oj}}{L + 2L_{oj}} & 0 \\ 0 & -\frac{R}{L} \end{bmatrix} \quad (13)$$

$$A_2(t) = A_2' \text{diag}(u(t)) \quad (14)$$

where $A_2' = [A_{21}'\ A_{22}'] \in \mathbb{R}^{2\times2N}$; $A_{21}' \in \mathbb{R}^{2\times N}$ is presented in (15); $A_{22}' \in \mathbb{R}^{2\times N}$ is presented in (16); and $u(t)$ is the input control vector, presented in (17).

$$A_{21}' = \begin{bmatrix} \frac{-1}{L+2L_{oj}} & \frac{-1}{L+2L_{oj}} & \cdots & \frac{-1}{L+2L_{oj}} \\ -\frac{1}{2L} & -\frac{1}{2L} & \cdots & -\frac{1}{2L} \end{bmatrix} \quad (15)$$

$$A_{22}' = \begin{bmatrix} \frac{1}{L+2L_{oj}} & \frac{1}{L+2L_{oj}} & \cdots & \frac{1}{L+2L_{oj}} \\ -\frac{1}{2L} & -\frac{1}{2L} & \cdots & -\frac{1}{2L} \end{bmatrix} \quad (16)$$

$$u(t) = \begin{bmatrix} S_{ju,1}, S_{ju,2}, \cdots, S_{ju,N}, S_{jl,1}, S_{jl,2}, \cdots, S_{jl,N} \end{bmatrix} \quad (17)$$

$$A_3(t) = \text{diag}(u(t))A_3' \quad (18)$$

where $A_3' \in \mathbb{R}^{2N\times2}$ is as follows:

$$A_3' = \begin{bmatrix} \frac{1}{2C_{ju,1}} & \cdots & \frac{1}{2C_{ju,N}} & \frac{-1}{2C_{jl,1}} & \cdots & \frac{-1}{2C_{jl,N}} \\ \frac{1}{C_{ju,1}} & \cdots & \frac{1}{C_{ju,N}} & \frac{1}{C_{jl,1}} & \cdots & \frac{1}{C_{jl,N}} \end{bmatrix}^T \quad (19)$$

$$D = \begin{bmatrix} D_1 \\ 0 \end{bmatrix} \in \mathbb{R}^{(2N+2)\times2} \quad (20)$$

where D_1 is as follows:

$$D_1 = \begin{bmatrix} \frac{-2}{L+2L_{oj}} & 0 \\ 0 & \frac{1}{2L} \end{bmatrix} \quad (21)$$

The output voltage of the MMC, such as phase-j, is defined as the voltage difference from point j to N.

$$v_{jN} = R_{oj}i_{oj} + L_{oj}\frac{di_{oj}}{dt} + u_j \quad (22)$$

In (10), the control of the system is to adjust the structure of matrix A so that $u_j \to u_j^*$, $U_{dc} \to U_{dc}^*$ (or the active power and reactive power are close to their desired values). As the focus of the paper is the universal mathematical model of the controlled object, the control method is shown in reference [26], and will not be detailed here.

From Equation (8), we can get the formula of u_j as follows (in active inverter, $u_j \neq 0$, $R_{oj} = 0$; in passive inverter, $u_j = 0$, $R_{oj} \neq 0$):

$$u_j = -\frac{1}{2}(L+2L_{oj})\frac{di_{oj}}{dt} - \frac{R}{2}i_{oj} - \frac{1}{2}u_{ju} + \frac{1}{2}u_{jl} \quad (23)$$

3. Algorithm of the Universal Mathematical Model

Equation (10) shows that the MMC is a nonlinear multi input system where the nonlinearity consists of the products between the states and inputs. The direct solution is difficult to find; as a result, the discrete sampling method will be used. The algorithm of the universal model is shown in Figure 2.

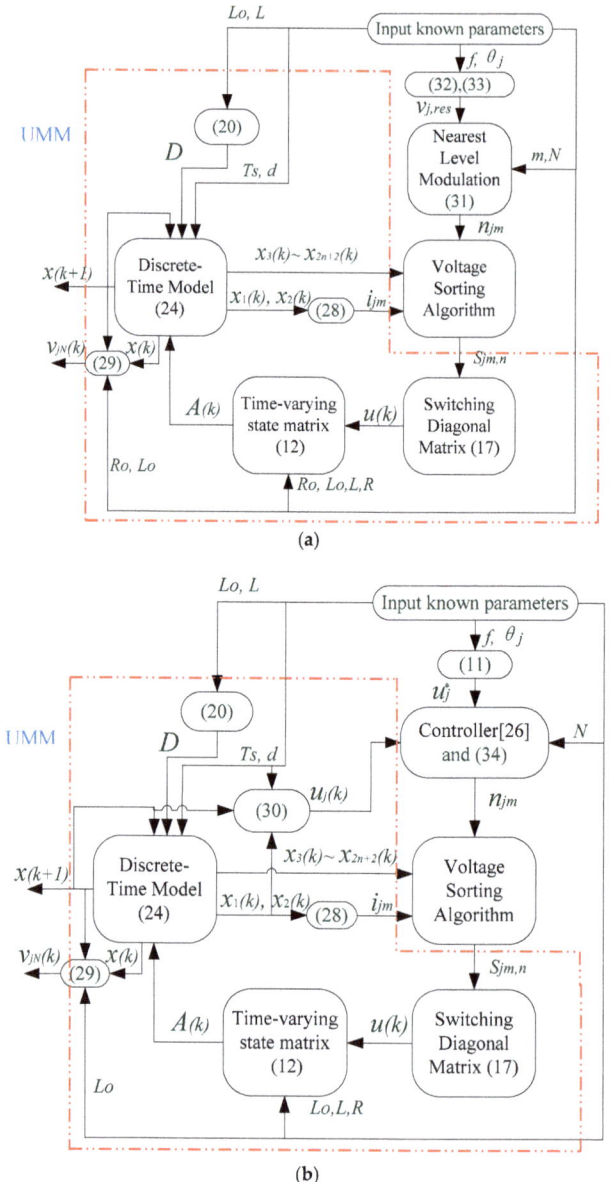

Figure 2. Block diagram of the universal mathematical model algorithm for MMC. (**a**) Open-loop control system, (**b**) closed-loop control system.

3.1. Discrete-Time Model of the MMC

It is assumed that the switching of equations occurs at the sampling points. Based on (10) and assuming a sampling time of T_s, the discrete-time model of the MMC, based on a forward Euler approximation, is obtained as:

$$x(k+1) = (I + T_s A(k))x(k) + T_s D d(k) \qquad (24)$$

The output current, $i_{oj}(k)$, can be calculated as:

$$i_{oj}(k) = x_1(k) \tag{25}$$

The circulating current, $i_{cj}(k)$, is obtained by Equation (26):

$$i_{cj}(k) = x_2(k) \tag{26}$$

The capacitance voltage, $u_{jm,n}(k)$, is obtained by the following Equation:

$$u_{jm,n}(k): x_3(k) \sim x_{2N+2}(k) \tag{27}$$

Based on Equations (6), (25), and (26), the arm currents can be calculated as:

$$i_{ju}(k) = \frac{x_1(k)}{2} + x_2(k), \quad i_{jl}(k) = -\frac{x_1(k)}{2} + x_2(k) \tag{28}$$

Based on Equation (22), the output voltage of the MMC can be expressed as:

$$v_{jN}(k) = R_{oj}x_1(k) + L_{oj}\frac{x_1(k+1) - x_1(k)}{T_s} + u_j(k) \tag{29}$$

Based on Equation (23), the AC voltage of the MMC can be expressed as:

$$u_j(k) = -\frac{1}{2}(L + 2L_{oj})\frac{x_1(k+1) - x_1(k)}{T_s} - \frac{R}{2}x_1(k) - \frac{1}{2}u_{ju}(k) + \frac{1}{2}u_{jl}(k) \tag{30}$$

3.2. Nearest Level Modulation

The nearest level modulation (NLM), also known as the round method, is an approach that uses the nearest voltage level to estimate the desired output voltage. The three phases are controlled independently. Given a normalized voltage reference $v_{j,res}$, the nearest output voltage level n_{jm} can be determined by:

$$\begin{cases} n_{ju} = \frac{N}{2} - \text{round}\left(\frac{mNv_{j,res}}{2}\right) \\ n_{jl} = \frac{N}{2} + \text{round}\left(\frac{mNv_{j,res}}{2}\right) \end{cases} \tag{31}$$

where m is the modulation coefficient, and $v_{j,res}$ is defined as:

$$v_{j,res} = \sin(2\pi ft + \theta_j) \tag{32}$$

where θ_j is the initial phase angle in phase-j. Normalized voltage references for the three phases can be presented as:

$$\begin{cases} v_{a,res} = \sin(2\pi ft) \\ v_{b,res} = \sin\left(2\pi ft - \frac{2\pi}{3}\right) \\ v_{c,res} = \sin\left(2\pi ft + \frac{2\pi}{3}\right) \end{cases} \tag{33}$$

3.3. Control Systems

In this paper, the control strategy applied in [26] is utilized to obtain the optimal value u_{ju}^* and u_{jl}^* through the tracking control of the AC voltage or DC voltage. Of course, not limited to this control strategy, other controls (such as the traditional PI control) are also applicable.

$$\begin{cases} n_{ju} = \text{round}\left(\dfrac{u_{ju}^*}{u_C}\right) \\ n_{jl} = \text{round}\left(\dfrac{u_{jl}^*}{u_C}\right) \end{cases} \quad (34)$$

where u_C is the capacitor-rated voltage, $u_C = U_{dc}^*/N$.

The MMC control diagram is shown in Figure 3, and a more detailed closed-loop control is shown in Figure 2b. Since this study focuses on the universality of the model controlled (UMM), only the system is considered as an open-loop system (in Figure 2a) in the early stage of the design, and the control system is designed after the system-controlled parameters are fixed.

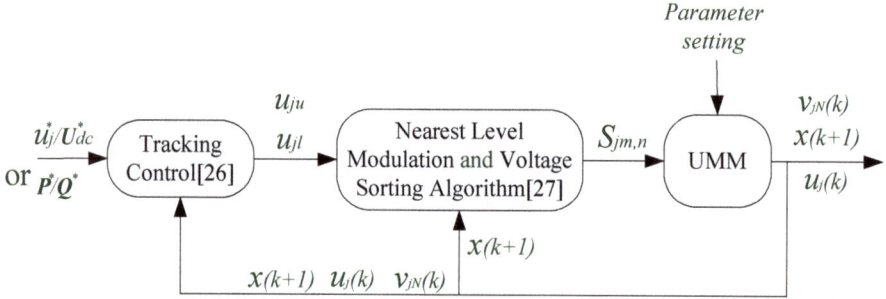

Figure 3. MMC control diagram.

3.4. Voltage Sorting Algorithm

In this paper, the voltage sorting algorithm applied in [27] is utilized to equalize all the capacitor voltages of the MMC $u_{jm,n}$. The algorithm reads the insertion indices n_{ju} and n_{jl} and determines which SMs are connected or bypassed in each arm of the MMC according to the plus-minus of the arm current i_{jm}. For example, if $i_{jm}(k) > 0$, the algorithm connects n_{jm} SMs with the lowest voltages in the corresponding arm and bypasses all the others. Conversely, if $i_{jm}(k) < 0$, the algorithm connects n_{jm} SMs with the highest voltages and bypasses the others. Therefore, the switching signals $S_{jm,n}$ to be applied in the sampling time k can be obtained. Finally, $u(k)$ is obtained based on (17).

4. Verification of Universal Mathematical Model

To evaluate the performance of the proposed UMM, a comparison between the results from the UMM and the nonlinear time-domain simulation model has been conducted. The nonlinear time-domain simulation model (SIM) is implemented in MATLAB/Simulink, and the UMM is performed using an m-file in MATLAB. The initial value is set as $x(0) = [0, 0, u_C, u_C, \ldots, u_C]^T \in \mathbb{R}^{2N+2}$. The comparison was conducted for a single-phase converter with 20 submodules. The main parameters of the MMC in the simulation are listed in Table 1. All the simulations were conducted using a Microsoft Windows 10 operating system with a 2.7 GHz Intel® core ™ i7-7500U processor and 16 GB of RAM. The test results are given in the following sub-sections.

Table 1. Main parameters of the MMC (phase-a).

Parameters	Value
AC system frequency f (Hz)	50
DC voltage U_{dc} (kV)	60
Initial capacitor voltage u_C (V)	U_{dc}/N
SM number in each arm (N)	20 (variable)
SM capacitance C (mF)	40 (variable)
Arm inductance L (mH)	3 (variable)
Arm equivalent resistance R (Ω)	0.5
Load inductance L_{oa} (mH)	400
Load resistance R_{oa} (Ω)	500
Sampling time T_s (μs)	50
AC side voltage v_a RMS (V)	0 (passive network)

4.1. Accuracy Analysis

The waveforms calculated by the UMM were compared with those generated from the SIM to evaluate the accuracy of the proposed approach. The results are shown in Figures 4 and 5. In these figures, the solid line "I" represents the result of UMM, the dashed line "II" represents the result of SIM.

4.1.1. Dynamic Simulation under Normal Operation

Figure 4 displays the out current, i_{oa}, and output voltage, v_{aN}, versus time. It shows that within a 0.2 s time range, the results calculated via UMM are favorable compared with the result obtained from SIM. According to Table 2, the root mean square errors are within a reasonable range. These results demonstrate the accuracy of the UMM method at steady state.

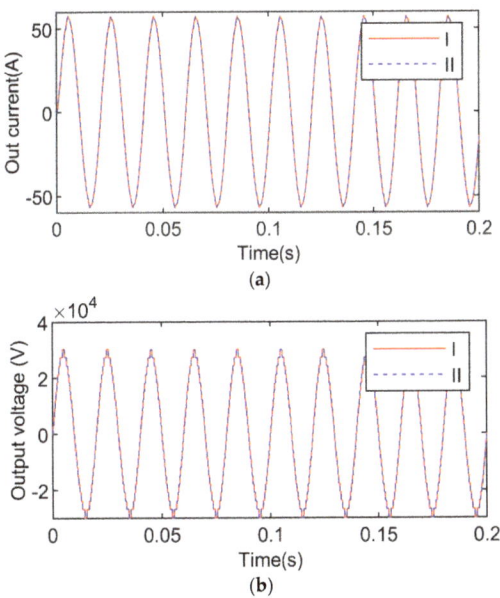

Figure 4. Output current/voltage diagram of the MMC. "I" represents the result of UMM, "II" represents the result of SIM. (a) current i_{oa}, (b) voltage v_{aN}.

For example, Figure 5a shows the transient behaviors of the circulating current. Figure 5b shows a trend of capacitor voltages for the upper and lower arm from transient to steady state. The upper arm current waveform is shown in Figure 5c. Before 0.06 s, the system was at a transient state, and the UMM and SIM results had shown similar behavior with a slight amplitude difference. After the transient state, the system tended to be stable and the two results were completely coincident.

The root mean square errors are shown in Table 2. From the numerical point of view, except for v_{aN} the errors of other variables are small. In the meantime, the value of v_{aN} is only less than 3/10,000 relative to its RMS (21,216 V).

In short, the perfect coincidence of the above-mentioned various dynamic waveforms (output current i_{oa}/voltage v_{aN}, circulating current i_{ca}, capacitive voltage $u_{am,n}$, and upper arm current i_{au}) has proved the accuracy and correctness of the proposed algorithm UMM.

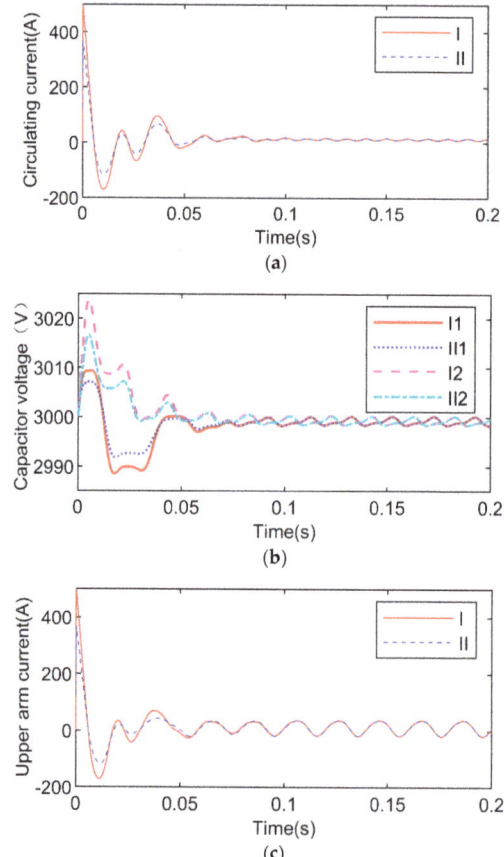

Figure 5. Circulating current/capacitor voltage diagram of the MMC. "I" represents the result of UMM, "II" represents the result of SIM. (a) circulating current i_{ca}, (b) voltage v_{aN}. "1" represents the upper arm, "2" represents the lower arm, (c) upper arm current i_{au}.

Table 2. Root mean square error of UMM and SIM.

Parameters	Root Mean Square Error	Parameters	Root Mean Square Error
i_{oa}	0.0061 A	i_{au}	0.0638 A
v_{aN}	6.4867 V	$v_{au,1}$	0.2855 V
i_{ca}	0.0668 A	$v_{al,1}$	0.6646 V

4.1.2. Dynamic Simulation of SM Open Circuit Fault

The open circuit fault of SM was assumed to study the effect of the internal fault of sub module on the MMC system. The time interval of failure is given as [3, 3.04]—i.e., two cycles.

In Figure 6, it can be seen that in the transient process of fault, the waveforms of UMM and that of SIM also match well.

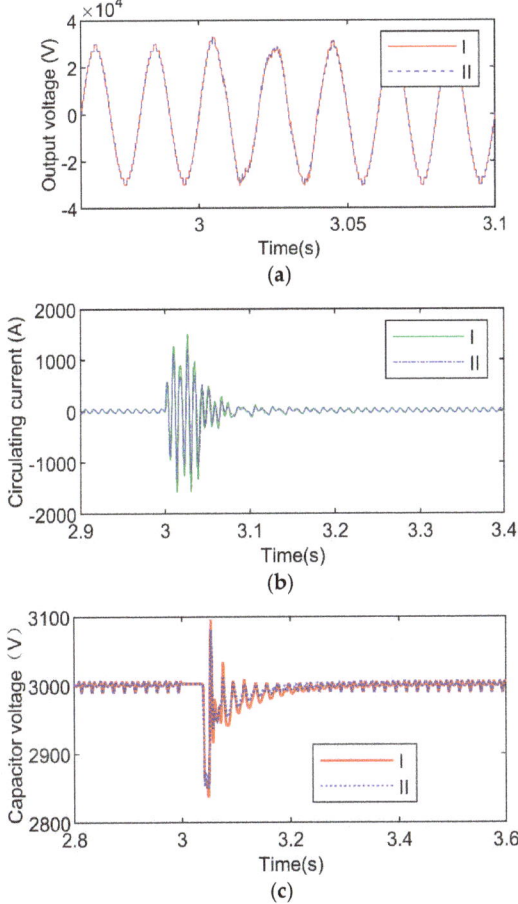

Figure 6. The waveform in case of open circuit fault. "I" represents the result of UMM, "II" represents the result of SIM. (**a**) output voltage, (**b**) circulating current, (**c**) capacitor voltage waveform in the case of open circuit fault.

4.2. Computational Efficiency

A 5 s period was tested with a 50 µs simulation time step. In this paper, we considered three groups of data (number of submodules, $N = 8, 12, 20$) for calculation, and the calculation results are shown in Figure 7. For example, when $N = 20$, the UMM took 7.1 s to get the result, while SIM consumed 62.2 s. The computational efficiency of UMM is significantly improved, which is about 8.7 times faster than that of SIM. However, the computational efficiency of the average value model (AVM) lies in the middle of the three.

Another advantage of UMM over SIM is that the SIM system took a large amount of time to build a system with many SMs.

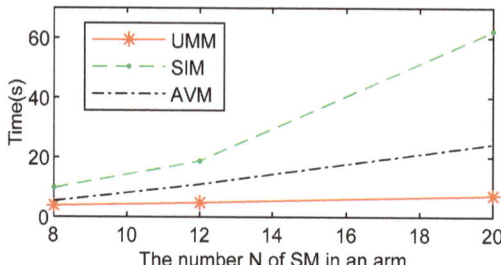

Figure 7. CPU times for UMM and SIM.

Under different sampling times (10, 30, 50 µs), the Central Processing Unit (CPU) times for UMM have been shown in Figure 8 as a 0.1 s period. It can be seen in Figure 8 that at $N < 150$, the three sampling times have little difference. However, with the increase in N, the time gap among them becomes larger. Especially when N is 404, the CPU times of the three were 162.8, 44.0, and 25.5 s, respectively. However, in fact, only the sampling time 50 µs for the UMM system met the accuracy requirements.

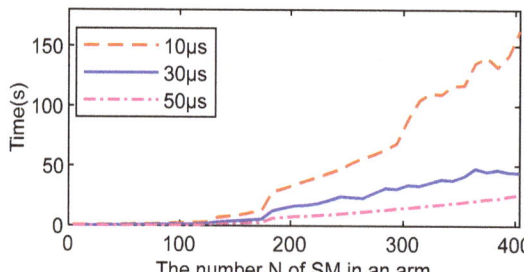

Figure 8. CPU times for UMM under different sampling times.

5. Application Based on UMM

Section 4 had proved the accuracy and efficiency of UMM. However, in this section, UMM will be used to study the MMC system, which is also the biggest difference from most existing models. The quality of the output voltage and current is a criterion to judge the performance of an inverter. For a long time, it was believed that the output current and voltage quality would be improved with an increase in the SM quantity, but there is no definite conclusion and mathematical evidence for this.

It is time-consuming to build simulation models, so it is impossible to simulate and analyze a large number of SMs. The utilization of UMM can facilitate this analysis. In this section, the influence of the number N of SMs on the output voltage and current THD under different conditions will be investigated. The dynamic performance of the MMC system was analyzed on the effects of different Ns.

5.1. Output Voltage/Current Harmonic Performance under Different N Values

The calculation parameters are given with Table 1, except for the parameter N, where N is an independent variable from 0 to 400. The simulation results are shown in Figure 9.

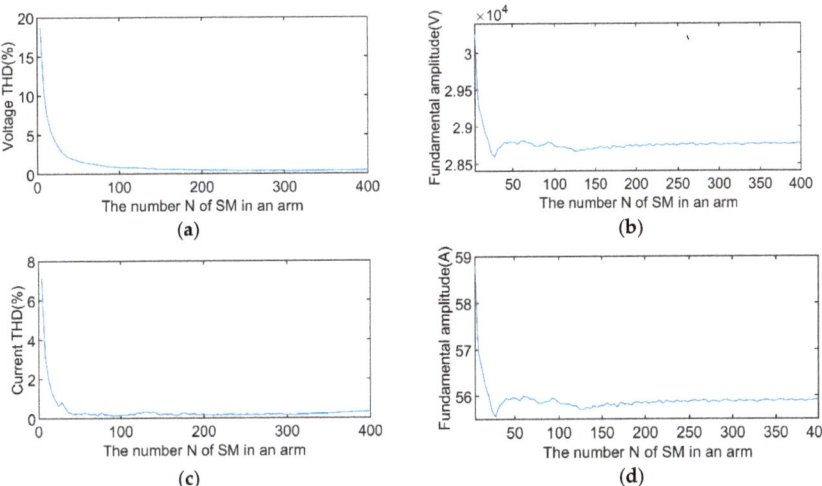

Figure 9. The oscillogram varying with the number of SM. (**a**) Output voltage THD, (**b**) output voltage fundamental amplitude, (**c**) output current THD, (**d**) output current fundamental amplitude.

In Figure 9, during the period when N was increased from 4 to 50, both the THD and fundamental amplitude decreased rapidly. After that, as N increased, the change rate tends to reduce. The optimal value was at (308, 0.352%) in Figure 9a, and (92, 0.1335%) in Figure 9c. The fundamental amplitude reached the minimum value at $N = 28$, then increased slightly in Figure 9b,d. After $N > 50$, it reached a stable state with small fluctuations.

In summary, the results show that the relationship between the quality of output voltage (or current) and the number N of SMs is not proportional. It has an optimal solution; using UMM, it is possible to find the optimal solution. In practice, in the design of N should be also considered the cost of hardware, the voltage grade, and the complexity of control.

5.2. Output Voltage/Current Harmonic Performance under Different SM Capacitance C and N Values

The influence of different SM capacitances, C, is studied in this section. The results are shown in Figure 10.

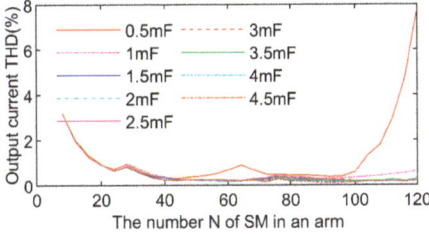

Figure 10. The THD diagram of the output current varying with C and N. The C value is shown in the legend.

Figure 10 displays the THD value of the output current and voltage changing with a different number N of SMs and SM capacitance C, respectively. As illustrated, when the N is small, less than 50, the THD basically coincides with different C values. After $N > 100$, the THD of the curve of $C = 0.5$ mF rises rapidly and become unstable. This phenomenon indicates that, along with the increase in the SMs, the capacity should be increased accordingly to keep the system stable.

According to [28], the voltage fluctuation rate of the SM capacitor can be calculated in the following equation:

$$\varepsilon = \frac{1}{3} \frac{S_{vN}}{N\omega C u_C^2} \quad (35)$$

where S_{vN} is the MMC nominal power, and $\omega = 2\pi f$.

Substitute $u_C = U_{dc}/N$ into (32), then:

$$\varepsilon = \frac{N S_{vN}}{3\omega C U_{dc}^2} \quad (36)$$

Generally, S_{vN}, ω, and U_{dc} are constants. As a result, ε is directly proportional to N and inversely proportional to C. When N increases and C is a fixed value, the system will diverge and become unstable after ε exceeds the allowable value.

5.3. Output Voltage/Current Harmonic Performance under Different Arm Inductance L and N Values

The influence of different SM inductances, L, is studied in this section. The simulation results are shown in Figure 11.

In Figure 11, when N is less than 110, the THD of the output voltage is basically the same for all the inductance values; after that, the THD decreases slowly as N increases, but the larger the arm inductance value, the higher the THD. It shows a small mutation near $N = 140$.

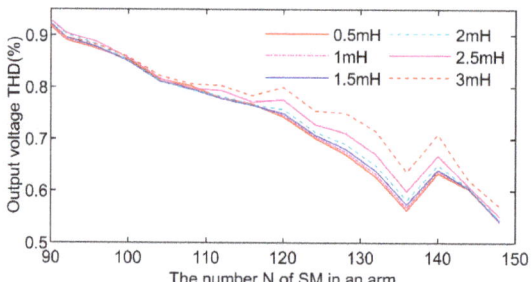

Figure 11. The THD diagram of the output voltage varying with L and N. The L value is shown in the legend.

Figure 12 shows the THD value of the output voltage changing with N and L. When N is less than 40, the output current THD is around the same value. The inductance value has different effects on THD in different areas of N. For example, the areas with the smallest THD are as follows: $L = 1$ mF, $N \in [44, 56]$; $L = 1.5$ mF, $N \in [56, 68] \cap [100, 116]$; $L = 3$ mF in the interval $[68, 80]$, $L = 0.5$ mF, $N \in [68, 100] \cap [116, 144]$ in the interval $[68, 100]$. After N increased to greater than 96, the THD increases significantly with the increase in L.

From the general trend, as N becomes larger, the inductance value should become smaller. However, if the inductance value is too small, it will lose the effect of suppressing the arm current and fault tolerance. Therefore, the design should be selected based on the actual situation.

5.4. Dynamic Performance under SM Capacitance Decay Fault

In this test, the capacitance of the SM_{au1} was attenuated by 0.6 times. Shown as Figure 13, the failure occurred at 3 s and continued until the end. From the SM voltage waveform, the difference between the waveforms before and after 3 s is obvious. This means that, in addition to the IGBT open circuit fault mentioned above, the proposed model can also study the parameter fault.

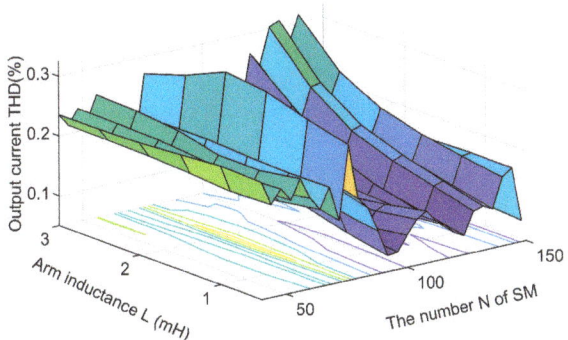

Figure 12. The THD diagram of the output current varying with L and N.

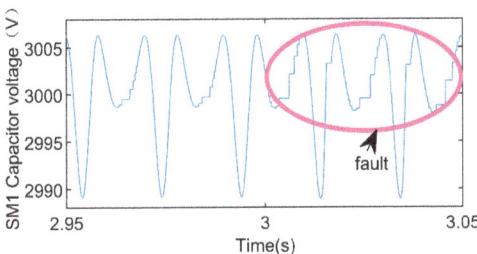

Figure 13. The SM1 capacitor voltage diagram under capacitance decay fault.

5.5. Object-Oriented Parametric Design

Through object-oriented programming in Figure 14, the program can realize human-computer interaction and reduce the workload of designers, so as to realize the universality, versatility, ease of use, and operation of MMC. Of course, this is only an example. See reference [28] for the calculation of relevant parameters, which will not be discussed here.

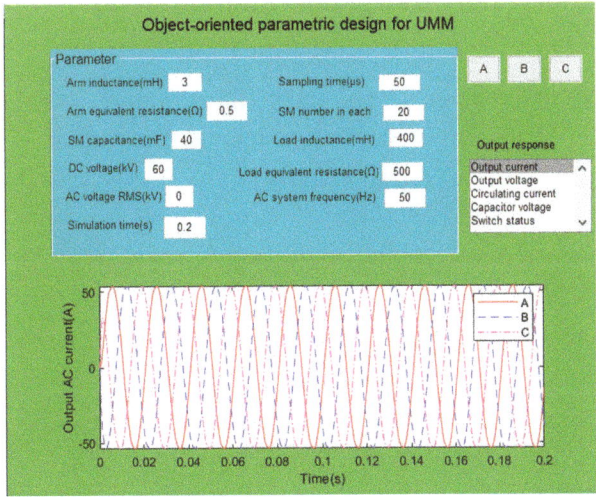

Figure 14. Design and verification interface for MMC.

6. Conclusions

A general mathematical model has been derived in detail from the circuit structure of MMC in this paper. The mathematical model includes the non-linear or linear characteristics of switching state, capacitance voltage, inductance, and resistance. Compared with the traditional MATLAB/Simulink (21-level MMC), the output voltage/current, circulating current, and capacitor voltage coincide, and the accuracy of the proposed model is fully proved. Additionally, this model is 8.7 ($N = 20$) times faster than the traditional simulation method. By changing the arm induction, SM capacitance, and the number of SMs to study the impact of each parameter on the output current/voltage THD, the optimal number of SMs has been found. In addition, the proposed model can also analyze the structure and parameter faults of MMC. The UMM proposed lays a solid foundation for further research into the dynamic performances of MMC with a large number of SMs.

The UMM mainly analyzes the static/dynamic system by discretizing the state equation of MMC. In essence, it belongs to the analytical analysis mathematical model. Its iterative calculation speed is much faster than that of the simulation model. At the same time, it also reduces the modeling time of the large-scale component modules and improves the universality. This is the reason that although the simulation model compared is only based on MATLAB/Simulink, it is not lost generality.

Author Contributions: Conceptualization, M.L. and Z.L.; methodology, M.L.; software, M.L.; validation, M.L.; resources, X.Y.; data curation, X.Y.; writing—original draft preparation, M.L.; writing—review and editing, M.L.; project administration, X.Y.; funding acquisition, Z.L. All authors have read and agreed to the published version of the manuscript.

Funding: This research was funded by the National Natural Science Foundation of China (No. 61963009) and (No. 61861007); Science and Technology Planning Project of Guizhou Province (No.2154 (2019)) and (No. 2302 (2016)); Collaborative Foundation of Guizhou Province (No. 7228 (2017)); Platform Talent Project of Guizhou Province (No. 5788 (2017)); and the special fund project of provincial governor for outstanding science and technology education talents in Guizhou Province (No. 4 (2010)).

Conflicts of Interest: The authors declare no conflict of interest.

Abbreviations

MMC	modular multilevel converter
THD	total harmonic distortion
UMM	universal mathematical model
SIM	simulation model
NLM	nearest level modulation
TDM	traditional detailed model
AM	accelerated model
DP	dynamic phasor
HVDC	High-voltage direct voltage
SM	sub-module
DC	direct current
AC	alternating current
IGBT	insulated gate bipolar transistor
AVM	average value model
EAM	enhanced accelerated model
EMT	electromagnetic transient

References

1. Raju, M.N.; Sreedevi, J.; Mandi, R.P.; Meera, K.S. Modular multilevel converters technology: A comprehensive study on its topologies, modelling, control and applications. *IET Power Electron.* **2019**, *12*, 149–169. [CrossRef]
2. Kouro, S.; Malinowski, M.; Gopakumar, K.; Pou, J.; Franquelo, L.G.; Wu, B.; Rodriguez, J.; Perez, M.A.; Leon, J.I. Recent advances and industrial applications of multilevel converters. *IEEE Trans. Ind. Electron.* **2010**, *57*, 2553–2580. [CrossRef]

3. Debnath, S.; Member, S.; Qin, J.; Member, S.; Bahrani, B. Operation, Control, and Applications of the Modular Multilevel Converter: A Review. *IEEE Trans. Power Electron.* **2015**, *30*, 37–53. [CrossRef]
4. Nguyen, T.H.; Al Hosani, K.; El Moursi, M.S.; Blaabjerg, F. An Overview of Modular Multilevel Converters in HVDC Transmission Systems with STATCOM Operation during Pole-to-Pole DC Short Circuits. *IEEE Trans. Power Electron.* **2019**, *34*, 4137–4160. [CrossRef]
5. Picas, R.; Zaragoza, J.; Pou, J.; Ceballos, S.; Konstantinou, G.; Capella, G.J. Study and comparison of discontinuous modulation for modular multilevel converters in motor drive applications. *IEEE Trans. Ind. Electron.* **2019**, *66*, 2376–2386. [CrossRef]
6. Li, B.; Zhou, S.; Xu, D.; Yang, R.; Xu, D.; Buccella, C.; Cecati, C. An Improved Circulating Current Injection Method for Modular Multilevel Converters in Variable-Speed Drives. *IEEE Trans. Ind. Electron.* **2016**, *63*, 7215–7225. [CrossRef]
7. Perez, M.A.; Bernet, S.; Rodriguez, J.; Kouro, S.; Lizana, R. Circuit topologies, modeling, control schemes, and applications of modular multilevel converters. *IEEE Trans. Power Electron.* **2015**, *30*, 4–17. [CrossRef]
8. Li, J.; Konstantinou, G.; Member, S.; Wickramasinghe, H.R.; Member, S.; Pou, J. Operation and Control Methods of Modular Multilevel Converters in Unbalanced AC Grids: A Review. *IEEE J. Emerg. Sel. Top. Power Electron.* **2019**, *7*, 1258–1271. [CrossRef]
9. Ilves, K.; Antonopoulos, A.; Norrga, S.; Nee, H.-P. Steady-state analysis of interaction between harmonic components of arm and line quantities of modular multilevel converters. *IEEE Trans. Power Electron.* **2012**, *27*, 57–68. [CrossRef]
10. Vatani, M.; Member, S.; Hovd, M.; Member, S. Control of the Modular Multilevel Converter Based on a Discrete-Time Bilinear Model Using the Sum of Squares Decomposition Method. *IEEE Trans. Power Del.* **2015**, *30*, 2179–2188.
11. Dekka, A.; Wu, B.; Fuentes, R.; Perez, M.; Zargari, N. Evolution of Topologies, Modeling, Control Schemes, and Applications of Modular Multilevel Converters. *IEEE J. Emerg. Sel. Top. Power Electron.* **2017**, *5*, 1631–1656. [CrossRef]
12. Vatani, M.; Member, S.; Bahrani, B. Indirect Finite Control Set Model Predictive Control of Modular Multilevel Converters. *IEEE Trans. Smart Grid* **2015**, *6*, 1520–1529. [CrossRef]
13. Wang, J.; Liang, J.; Gao, F.; Xiaoming, D.; Wang, C.; Zhao, B. A Closed-Loop Time-Domain Analysis Method for Modular Multilevel Converter. *IEEE Trans. Power Electron.* **2017**, *32*, 7494–7508. [CrossRef]
14. Saad, H.; Peralta, J.; Dennetiere, S.; Mahseredjian, J.; Jatskevich, J.; Martinez, J.A.; Davoudi, A.; Saeedifard, M.; Sood, V.; Wang, X. Dynamic averaged and simplified models for MMC based HVDC transmission systems. *IEEE Trans. Power Del.* **2013**, *28*, 1723–1730.
15. Leon, A.E.; Amodeo, S.J. Modeling, control, and reduced-order representation of modular multilevel converters. *Electr. Power Syst. Res.* **2018**, *163*, 196–210. [CrossRef]
16. Belhaouane, M.M.; Ayari, M.; Guillaud, X.; Braiek, N.B. Robust Control Design of MMC-HVDC Systems Using Multivariable Optimal Guaranteed. *IEEE Trans. Ind. Appl.* **2019**, *55*, 2952–2963. [CrossRef]
17. Xu, J.; Gole, A.M.; Zhao, C. The use of averaged-value model of modular multilevel converter in DC grid. *IEEE Trans. Power Del.* **2015**, *30*, 519–528.
18. Meng, X.; Han, J.; Bieber, L.; Wang, L.; Li, W.; Belanger, J. A Universal Blocking-Module-Based Average Value Model of Modular Multilevel Converters with Different Types of Submodules. *IEEE Trans. Energy Convers.* **2020**, *35*, 53–66. [CrossRef]
19. Lyu, J.; Zhang, X.; Cai, X.; Molinas, M. Harmonic State-Space Based Small-Signal Impedance Modeling of a Modular Multilevel Converter with Consideration of Internal Harmonic Dynamics. *IEEE Trans. Power Electron.* **2019**, *34*, 2134–2148. [CrossRef]
20. Freitas, C.M.; Watanabe, E.H.; Monterio, L.F.C. A linearized small-signal Thévenin-equivalent model of a voltage-controlled modular multilevel converter. *Electr. Power Syst. Res.* **2020**, *182*, 106231–106241. [CrossRef]
21. Gnanarathna, U.N.; Gole, A.M.; Jayasinghe, R.P. Efficient modeling of modular multilevel HVDC converters (MMC) on electromagnetic transient simulation programs. *IEEE Trans. Power Del.* **2011**, *26*, 316–324.
22. Xu, J.; Zhao, C.; Liu, W.; Guo, C. Accelerated model of modular multilevel converters in PSCAD/EMTDC. *IEEE Trans. Power Del.* **2013**, *28*, 129–136.
23. Beddard, A.; Barnes, M.; Preece, R. Comparison of Detailed Modeling Techniques for MMC Employed on VSC-HVDC Schemes. *IEEE Trans. Power Deliv.* **2015**, *30*, 579–589.

24. Rupasinghe, J.; Member, S.; Filizadeh, S.; Member, S. A Dynamic Phasor Model of an MMC with Extended Frequency Range for EMT Simulations. *IEEE J. Emerg. Sel. Top. Power Electron.* **2019**, *7*, 30–40. [CrossRef]
25. Rashidi, N.; Burgos, R.; Roy, C.; Boroyevich, D. On the Modeling and Design of Modular Multilevel Converters with Parametric and Model-Form Uncertainty Quantification. *IEEE Trans. Power Electron.* **2020**, *35*, 10168–10179. [CrossRef]
26. Liu, M.; Li, Z.; Yang, X. Tracking Control of Modular Multilevel Converter Based on Linear Matrix Inequality without Coordinate Transformation. *Energies* **2020**, *13*, 1978. [CrossRef]
27. Gutierrez, B. Modular Multilevel Converters (MMCs) Controlled by Model Predictive Control with Reduced Calculation Burden. *IEEE Trans. Power Electron.* **2018**, *33*, 9176–9187. [CrossRef]
28. Zheng, X. *HVDC System*, 2nd ed.; China Machine Press: Beijing, China, 2016; pp. 69–71.

© 2020 by the authors. Licensee MDPI, Basel, Switzerland. This article is an open access article distributed under the terms and conditions of the Creative Commons Attribution (CC BY) license (http://creativecommons.org/licenses/by/4.0/).

Article

Inrush Current Control of High Power Density DC–DC Converter

Ahmed H. Okilly [1], Namhun Kim [2] and Jeihoon Baek [1],*

- [1] Electrical & Electronics and Communication Engineering Department, Koreatech University, Cheonan-si 31253, Korea; ahmed21490@koreatech.ac.kr
- [2] Research Center, ESTRA Automotive, Daegu-si 42981, Korea; namhun.kim@estra-automotive.com
- * Correspondence: jhbaek@koreatech.ac.kr; Tel.: +82-(04)-15601258

Received: 29 June 2020; Accepted: 18 August 2020; Published: 19 August 2020

Abstract: This paper presents a complete mathematical design of the main components of 2 kW, 54 direct current (DC)–DC converter stage, which can be used as the second stage of the two stages of alternating current (AC)–DC telecom power supply. In this paper, a simple inrush current controlling circuit to eliminate the high inrush current, which is generated due to high input capacitor at the input side of the DC–DC converter, is proposed, designed, and briefly discussed. The proposed circuit is very easy to implement in the lab using a single metal–oxide–semiconductor field-effect transistor (MOSFET) switch and some small passive elements. PSIM simulation has been used to test the power supply performance using the value of the designed components. Furthermore, the experimental setup of the designed power supply with inrush current control is built in the lab to show the practical performance of the designed power supply and to test the reliability of the proposed inrush current mitigation circuit to eliminate the high inrush current at initial power application to the power supply circuit. DC–DC power supply with phase shift zero voltage switching (ZVS) technique is chosen and designed due to its availability to achieve ZVS over the full load range at the primary side of the power supply, which reduces switching losses and offers high conversion efficiency. High power density DC–DC converter stage with smooth current startup operation, full load efficiency over 95%, and better voltage regulation is achieved in this work.

Keywords: DC–DC converter; phase shift PWM; ZVS; inrush current; MOSFET; telecom server

1. Introduction

The spread of 5G technology in communications and telecom systems making universal electronic devices makes 5G technology one of the largest sources of electrical energy consumption, including such electronic devices usually operated with DC power, so that the supply AC voltage needs to be rectified. Conventional AC–DC rectifiers can be used to supply these devices, but circuit performance and power efficiency at high power density applications limit the use of such rectifiers; therefore, their energy efficiency must be increased. Recently, highly efficient AC–DC power supply with a high power factor has been modified for this purpose. Figure 1 illustrates the general construction of the AC–DC telecom power supply, where the two stages power supply consists of the power factor correction (PFC) stage and where the DC–DC output converter stage is the best option to get high power performance and good energy quality [1–3].

The design of power supply with a high input power factor requires the modifying of the input AC current waveform to follow the input voltage waveform to reduce the phase shaft between them, and then the reduction of the harmonic contents of the input current [3]. The power supply first stage usually includes an electromagnetic interference (EMI) filter, bridge rectifying circuit, boost converter, and output bulk capacitor; by controlling the boost converter operation, the input power factor can approach to, or near to, unity [4]. Single switch, two switches or full bridge switching topology can be used to implement PFC boost converter, but switching topology proper switches must be used to withstand the output voltage stress of this stage normally (320–400 V) DC [1]. Different techniques and topologies of the analog control PFC stage of the telecom power supply with power factor more than 99% are previously discussed in [5].

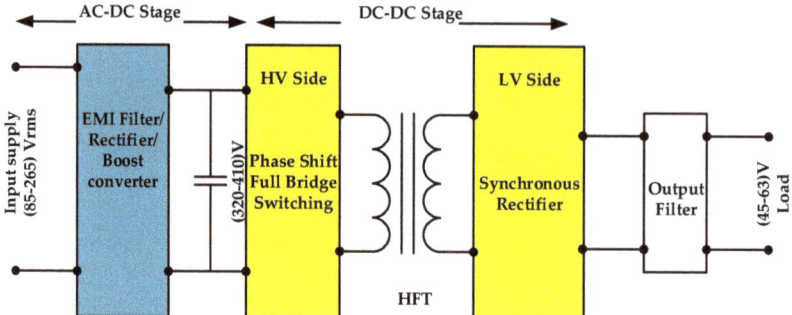

Figure 1. Block diagram of the complete AC–DC power supply for a telecom server.

The power supply second stage is the target of this paper and it is designed to regulate the PFC output voltage to the required load voltage (usually 45–63 V) for the telecom applications. In this stage we must choose an appropriate technique to provide high power density conversion with high efficiency [6], DC–DC converter based on phase shifted ZVS technique is one of the most widely used techniques, because it has features to minimize the switching losses and offer better regulation over a wide load range [7]. Moreover, by using the ZVS technique, it is available to have high power density conversion with low voltage stress and small switching losses [8–10].

Switches on both sides of the high power density DC–DC converter must be designed to withstand the high voltage stress in the primary side and the high current stress in the secondary side [10]. So, one of the most important issue in designing DC–DC converter circuits is to choose the appropriate switching schematics for the converter to reduce the stress in the switches during the converter operation. Based on the phase shifted converter applications, phase shift pulse width modulation (PWM) converter can be implemented using different switching schematics (full bridge, half bridge, dual half bridge) in primary and secondary sides, and analysis of phase shifts isolated DC–DC converter different topologies previously given in [11]. Analysis and design of phase shift controller for a dual half bridge DC–DC converter is provided in [12]. Based on the application of DC–DC converter, it can be designed as an isolated or non-isolated converter; for the isolated DC–DC converter, transformer design and choice are a big challenge to have ZVS for the wide range of the load, which can reduce the switching and conduction losses [13]. Analysis of the phase shifted converter by using series connected transformers for low conduction losses are given in [14,15]. Design and implementation of the non-isolated phase shift ZVS converter is presented in [16].

DC–DC converter control circuits can be implemented using analog or digital control circuits, but analog control chip as compared with digital micro controller unit (MCU) has some demerits such as the temperature drift, fixed control parameters, and slow response speed; on the other hand the price of the MCU and the required analog digital interfacing sensors are expensive as compared with the analog control technique requirement. Optimization analysis and the design of different control techniques for the DC–DC converter have been previously presented in [17–21]. Design of 1 kW efficient phase shift telecom DC–DC converter based on the maximum duty cycle and optimal hold-up time is given in [22]. Controlling techniques to reduce the switching and conduction loss are presented in [23,24].

The bulk capacitor, which is introduced between the two stages of power supply as shown in Figure 1 for adjustment of the input voltage hold-up time and harmonic values of the input waveform, causes high inrush current for a few cycles where a very high "dv/dt" occurred at the initial power application to the power supply; this high dv/dt cause spikes of short-duration and a high peak current, which value may be higher than the circuit component rating current and can seriously damage or destroy these components [25–27]. So appropriate inrush current controlling circuit must be designed to limit such this current. Usually, limiting and reducing of the high inrush current is done by using a large size inductor or resistors in series with the input capacitors [28], but by using these techniques, converter compact design, and weight and power losses cannot be optimally utilized.

In order to reduce the power dissipation on the series resistor, a parallel semiconductor switch or relay can be connected through the resistance. However, based on the converter operating current, size of the relay can be also excessive large. Moreover, by using of a semiconductor switch an appropriate controlling circuit must be designed to control the switch operation. Another technique used to control the inrush current is by handling a soft starter time at the beginning of the DC–DC converter operation; soft starter technique works to limit inrush current based on controlling the duty cycle of the converter switches in order to slowly charge the capacitors, and hence reduce the high dv/dt for limiting the inrush current [29–31]; this technique is usually implemented to limit the inrush current inside the analog controller chips such as PWM controller integrated circuits (IC's) UCC256403 and UCC28950 from Texas Instruments, Dallas, Texas, United States, 2016 [32].

Recently, NTC thermistor is the most famous device used in reducing the inrush current generated in the telecom power supplies. An NTC thermistor has a compact size and a negative temperature coefficient, is connected in series with the input power supply, and when the current flows, its temperature will increase and the device resistance will decrease [33,34]. The shortcoming of this device is that it requires a cool off time after the stopping of the current flowing in order to return back the device high resistance.

Another more applicable and simple technique to limit the converter inrush current can be done using a single MOSFET switch connected with the input side of the DC–DC converter [35–37]. MOSFETs switches are usually considered as ideal devices because they are characterized by fast switching time due to majority carrier, lower switching losses due to fast rise and fall times, as well as very small on-state DC resistance, which helps to reduce the voltage drop through the switch at steady state operation [38]. Control of the inrush current to the required limit can be done by controlling the gate charge transfer characteristics of the MOSFET switch in order to control the slew rate of the input capacitance charging time [39]. This technique offers inrush current control with economic price, compact design, and simple implementation, without using any sophisticated control circuit for the MOSFET operation [38–40]. High reliability using this technique can be obtained with the appropriate choice of the MOSFET switch, as well as design of the biasing circuit schematic and the connection strategy to the input side of the DC–DC converter [41].

In this paper, complete design, analysis, and mathematical calculations of the circuit main components and control systems of 2 kW, 54 V telecom DC–DC converter stage with phase shift ZVS technique, using full bridge at primary and synchronous rectification at secondary side are presented. A sample inrush current controlling circuit based on using MOSFET switch is proposed and is inserted in the input side of the DC–DC converter, proposed controlling circuit schematics as well as design and choice of the circuit components are briefly discussed. The main advantages of the proposed circuit are that the controlling of the gate charge transfer characteristics of the MOSFET occurs without any contribution or connection with the DC–DC converter control circuit, which make the proposed circuit usable with analog and digital control converters, and has fast response and high reliability to control the inrush current to the required value as well as simple implementation for any converter with different input–output operation conditions. The designed main components and control system of the power supply circuit make the power supply output voltage performance follows the standard specifications IEC61000-3-3, which is required by telecom applications.

The result sections of this paper are organized as follow: first, simulation of the complete designed converter is performed using PSIM software to be sure that the reliability of the designed components enhances the required power supply performance. Then, an experimental setup of the designed converter with the proposed inrush current mitigation circuit is inserted between the power supply two stages is performed in the lab, which shows that the designed DC–DC converter with proposed inrush control circuit can achieve smooth startup operation at the input side, where efficiency is more than 95.5%, and with better voltage regulation at the output side of the converter.

2. Design Procedure of the DC–DC Stage of the Telecom Power Supply

Figure 2 shows the schematic circuit of the telecom DC–DC converter stage with the proposed inrush current control circuit connected at the input side of the DC–DC phase shift converter. In designing the high-power density power supply with high conversion efficiency, losses in this stage must be maintained at their lowest value; one of the highest efficiency conversion techniques at high power density is the full bridge phase shift converter, which can offer very small switching losses by means of the ZVS technique at the converter primary side, which leads to increase the conversion efficiency [7,42].

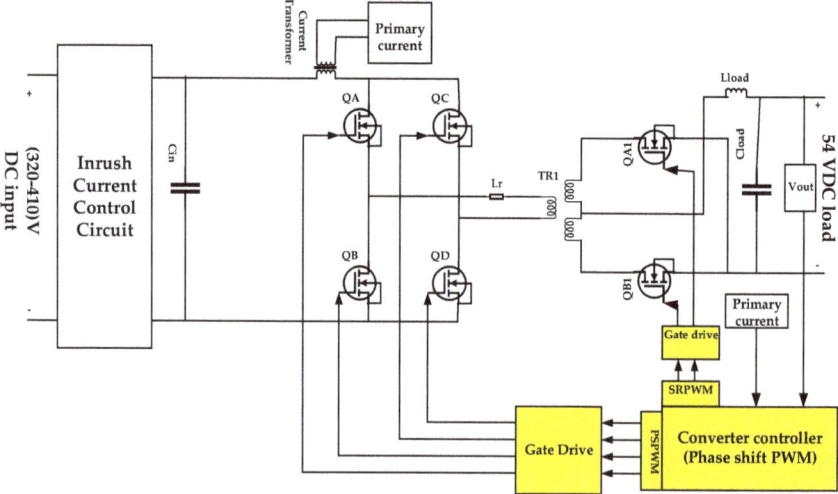

Figure 2. Schematic diagram of DC–DC phase shift ZVS converter with the proposed inrush current control.

The most important components in the DC–DC converter stage to be designed in the next subsections are the input filter capacitor (C_{in}), inrush current control circuit, resonant inductor (L_r), transformer (T_{R1}) turns ratio (a) and magnetizing reactance (L_m), and load inductance and capacitance (L_{load} and C_{load}). Other important factors are the choice of appropriate switches for the full bridge rectifier at the primary side and synchronous rectifier at the secondary side, which can withstand the high voltage stress at transformer primary and high current rated at secondary side [42].

2.1. Input Capacitor Design

Input capacitor of the DC–DC converter is designed to meet the hold-up time (t_{hold}) for the minimum input voltage ($V_{in\ min}$) applied to the converter circuit [15,40], and can be calculated as

$$C_{in} \geq \frac{2 P_o\, t_{hold}}{\eta \left(V_{in}^2 - V_{in\ min}^2\right)} \tag{1}$$

where P_o refers to the rated converter power, V_{in} refers to the Converter input voltage, and η is the designed efficiency of the DC–DC converter.

2.2. Proposed Inrush Current Control Circuit

In this subsection the proposed inrush current control circuit based on the controlling of the gate charge characteristic of the MOSFET switch was designed to be used with any DC–DC converter topology (analog or digital); the proposed circuit was tested in PSIM software and also was implemented in the lab to show the power supply two stages (PFC and DC–DC converter) practical current characteristic with, and without, using this circuit.

The input of DC–DC converter stage usually contains filter to adjust the total harmonics distortion (THD) of the input voltage to the DC–DC converter. Input filters usually consist of passive elements such as capacitor and inductor; when power is initially applied to the DC–DC converter circuit, high inrush current will flow due to high dv/dt of the filter capacitor as illustrated in Figure 3, which shows the converter input current curves at initial power application to the converter circuit.

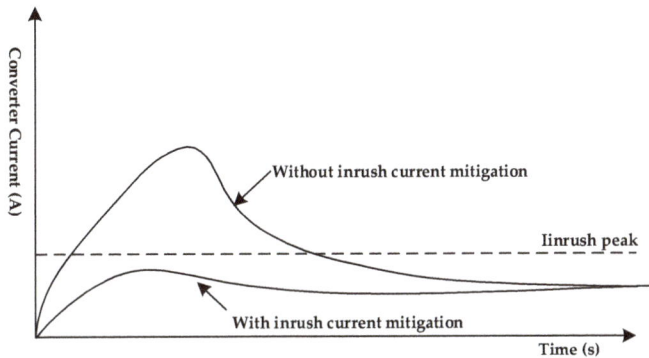

Figure 3. Inrush current at the startup operation of DC–DC converter.

Mathematical analysis, which briefly describes the startup response of the power supply operation and the phenomena of the inrush currents in DC–DC converters, are previously discussed in [26,27]. At the moment of applying the voltage to the DC–DC converter circuit, the current flowing through the input capacitor (C_{in}) can be expressed as

$$I_{Cin} = C_{in} \frac{dV_{in}}{dt} \tag{2}$$

This current has a high amplitude with very small duration and can damage or destroy circuit components if it exceeding the rating of the designed components. Therefore, this current must be effectively managed to ensure system safety operation and stability.

MOSFET switches are charge controlled devices, so if a MOSFET device is connected between two stages of the power supply, as shown in Figure 4, inrush current due to high dV_{in}/dt, which is caused by the input capacitor (C_{in}) at the initial power application, can be controlled by controlling the initially high dV_{in}/dt of the input capacitor by using the ability to control the constant linear slope of the drain voltage transition, which allows accurate control of the inrush current to the capacitive load. This is possible because the current flowing through the capacitor is dependent upon the transition of the voltage as shown in Equation (2). Reliability of this technique depends on the ability to control the MOSFET gate charge transfer curve illustrated in Figure 5 [29].

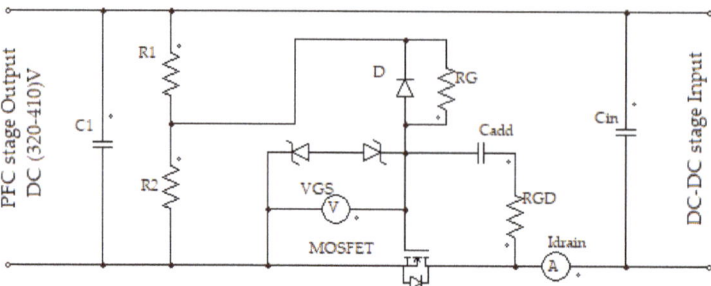

Figure 4. Proposed inrush current controlling circuit.

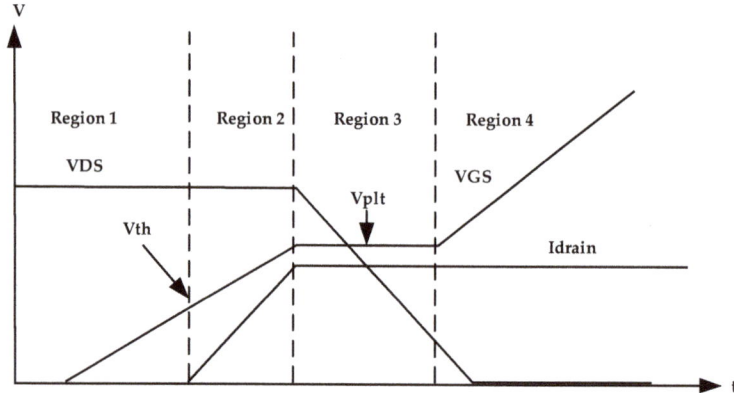

Figure 5. Gate charge transfer characteristics of MOSFET.

Figure 6 shows the equivalent representation of the MOSFET switch, where the gate charge curve is influenced by the MOSFET equivalent input capacitance [29].

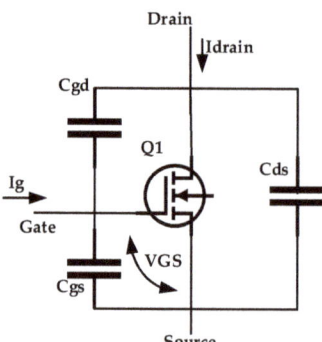

Figure 6. Circuit of N-MOSFET.

Cgd and Cgs represents the gate-drain and gate-source capacitances, respectively; these two capacitance values represent the input capacitance of the MOSFET device. Usually for MOSFET, Cgd is greater than Cgs. So Cgd strongly influences the input capacitance of the MOSFET, this means that MOSFET input capacitance can control the inrush current value, if the value of Cgd is controlled. By inserting the appropriate design capacitance, C_{add}, between the drain and gate terminals of MOSFET, as shown in the proposed circuit in Figure 4, Cgd can be controlled, which then can control the slew rate of the gate-source voltage, V_{GS}, which controls the slew rate of the input capacitance charging ($dVin/dt$), and consequently controls the inrush current of the circuit.

As shown in the characteristics in Figure 5, when MOSFET is turned on, the charging of the equivalent capacitance occurs at Region 1, and the charging time is determined by the equivalent input capacitance of MOSFET. The voltage (V_{GS}) increases until the starting point of Region 2, where it reaches the threshold value (V_{th}). At this time drain current starts to flow and the rate of increase of drain current given by

$$\frac{dI_{drain}}{dt} = g_f \frac{dV_{GS}}{dt} \tag{3}$$

where g_f is the switch forward trans-conductance and can be easily know from the switch datasheet.

At the end of Region 2, charging and discharging of Cgs simultaneously occurs, which causes V_{GS} to be maintained constant at Miller plat voltage (V_{plt}) as shown in Region 3. With V_{plt}, the drain current is saturated at peak constant value dependent of V_{GS} voltage value, and V_{plt} can be calculated as

$$V_{plt} = V_{th} + \frac{I_{drain}}{g_f} \tag{4}$$

The constant voltage of V_{GS} causes the input gate current to flow through the additional capacitance C_{add}, and can be calculated as

$$I_g = \frac{V_{GG} - V_{plt}}{R_G} = C_{add} \frac{dV_{DS}}{dt} \tag{5}$$

$$V_{GG} = V_{in} \times \frac{R_2}{R_1 + R_2} \tag{6}$$

where R_G is designed and connected in a series with the gate for controlling the gate current I_g. Finally, by controlling the value of V_{GS} and I_g, it is possible to control the maximum drain current of the MOSFET and control the inrush current at the initial power application to the power supply circuit, when the drain terminal of the MOSFET is connected to the return path of the DC–DC converter, as shown in Figure 4. R_{GD}, a small value resistor, as compared with R_G, connected with the C_{add} and prevents unwanted high frequency oscillation [29]. Using Equations (5) and (6) inrush peak current value is approximately given by

$$I_{inrush} = -I_{drain} = g_f \left(C_{add} \times R_G \times \frac{dV_{DS}}{dt} + V_{th} - V_{GG} \right) \tag{7}$$

The value of this current must be less than the maximum permissible DC current (inrush peak) of the primary side of the DC–DC converter stage.

Figure 7 shows the flow chart for the complete design procedure for the proposed inrush current control circuit by using the method of the controlling in the gate charge transfer characteristics of the MOSFET.

In Region 4, V_{GS} still increases to higher values. If this voltage reaches a value higher than the gate source breakdown voltage (BV_{GS}), MOSFET may be damaged; therefore, MOSFET must be protected at this region from the higher applied voltage; therefore, PFC output stage voltage (input voltage to the DC–DC converter) is divided using resistors R1 and R2 and only a small voltage is required to be applied to the MOSFET. Additionally, we can use one switch from the family of Zener-protected MOSFETs, such as switches from STMicroelectronics company [43]; in this protected switch, when the voltage applied to the switch is more than the breakdown voltage of the Zener (less than BV_{GS}), the Zener diode breaks down and the voltage is saturated at the safe limit. The design result of the active inrush current control circuit, which used in simulation and experimental modeling is shown in Table 1.

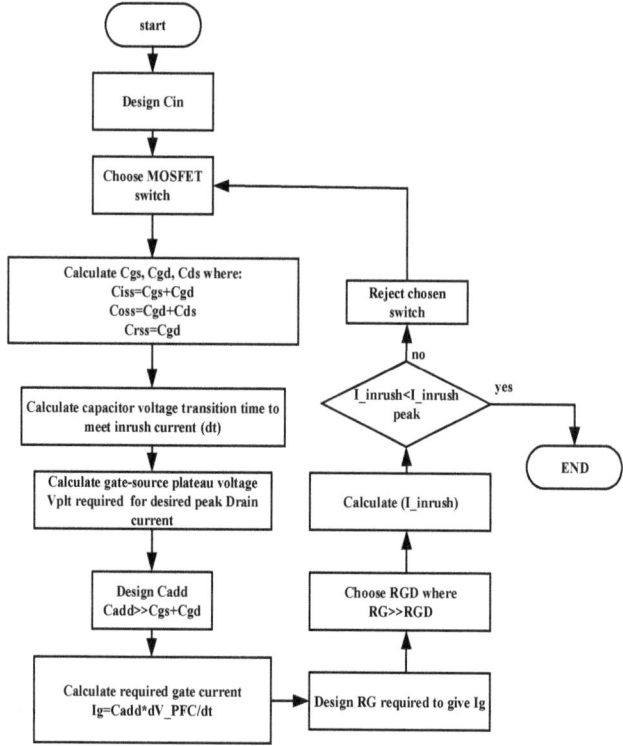

Figure 7. Flow chart of design process of the proposed inrush current control circuit.

Table 1. Design Results for Active Inrush Current Control Circuit.

Parameter	Value	Unit
Input capacitance (C_{in})	350	uF
Chosen MOSFET	STW20NK50Z	-
C_{add}	0.1	uF
R_G	30	kΩ
R_{GD}	500	Ω
R_1	715	kΩ
R_2	84.5	kΩ
Allowable Inrush peak	9	A
Plat voltage (V_{plt})	4.35	V
Gate current (I_g)	3.430	mA

2.3. Transformer (T_{R1}) Turns Ratio (a) and Magentizing Reactance (Lm) Calculation

Transformer turns ratio (a) is calculated based on maximum operating duty cycle (D_{max}) at the minimum input voltage rating of the converter ($V_{in\ min}$) as follows:

$$a = \frac{N_P}{N_S} = \frac{V_P}{V_S} \tag{8}$$

Let D_{max} be about 70% and $V_{in\ min}$ about 320 V, and the transformer turns ratio is calculated as

$$a = \frac{V_P}{V_S} = \frac{(V_{in\ min} - 2VD_Q)D_{max}}{V_{load} + VD_Q} = 4.1 \tag{9}$$

where VD_Q is the switch voltage drop and it is assumed to be 0.5 V in calculations.

Let $a = 5$, the typical operating duty cycle (D), is calculated as

$$D = \frac{(V_{load} + VD_Q)a}{(V_{in} - 2VD_Q)} = 0.68 \tag{10}$$

Transformer magnetizing inductance (L_m) designed based on the maximum magnetizing inductance to realize ZVS as expressed in [13,18], and can be expressed as

$$L_m = \frac{T_{dead}\ a\ V_{load\ min}}{C_{HB}\ V_{in\ min}} \times \left(\frac{T_{s\ min}}{4} - \frac{T_{dead}}{2}\right) \tag{11}$$

where C_{HB} refers to the total equivalent capacitance of the primary H bridge, which can be known from the primary switch data sheet. $T_{s\ min}$ is the minimum switching time depends on the designed minimum switching frequency, also T_{dead} is the PWM dead time which can be calculated according to the previously calculated duty ratio.

One more important issue is the choosing of a transformer with appropriate magnetizing inductance to minimize the output current ripple and to make sure that the converter works in the required control mode, where the smallest value of magnetizing inductance makes the converter work in voltage control mode instead of current control mode [10].

2.4. Resonant Inductor (Lr) Design

Resonant inductor tank is calculated based on the amount of energy required to achieve ZVS condition. The energy absorbed by the inductor values of the resonant inductance (L_r) and the transformer leakage inductance (L_{lk}) must be able to exhaust the energy supplied by the average parasitic capacitance of the primary switches (C_{ossavg}), and also the energy from the transformer winding capacitance (C_w) [42,44].

$$\frac{1}{2}I_p^2(L_r + L_{lk}) \geq \frac{4}{3}C_{ossavg}V_{in}^2 + \frac{1}{2}C_w V_{in}^2 \tag{12}$$

where I_p refers to the converter primary current (A).

2.5. Output Inductance and Capacitance (L_{load} and C_{load})

Output load inductor (L_{load}) is designed based on 10% ripple value in the load DC current (I_{load}), as follows:

$$\Delta I_{load} = \frac{P_{load} \times 0.10}{V_{load}} \tag{13}$$

$$L_{load} = \frac{V_{load} \times (1-D)}{\Delta I_{load} \times F_s} \tag{14}$$

Output load capacitor (C_{load}) is selected based on hold-up time (t_{hu}) and 20% (200 mV) of the allowable load transient voltage (V_{tran}) as follow:

$$C_{load} \geq \frac{0.9I_{load} \times t_{hu}}{0.2V_{tran}} \tag{15}$$

where hold-up time (t_{hu}) is calculated as the time required for the inductor current to reach to 90% of the full load current [41].

$$t_{hu} = \frac{L_{load} \times 0.9I_{load}}{V_{load}} \tag{16}$$

2.6. DC–DC Converter Controller Design and Implmentation

Phase shift PWM technique is used to control the full-bridge in the primary side of the DC–DC converter by phase shifting the switching pulses of one half-bridge with respect to the other. High power density efficient conversion is available using ZVS technique at high switching frequency in this part. Voltage-mode or current-mode control techniques can be used in this part. Current-mode controlled DC–DC switching is popular and provides a more highly efficient power conversion than voltage mode control. However, the current-mode design can suffer from instability when the duty cycle of the PWM rises above 50% [45]. To overcome this instability, converter primary current slope compensation technique is used to restore reliability over the wide range of duty-cycle [19,46].

Figure 8 shows the schematic of control technique of DC–DC converter, implemented in PSIM software where the primary current and the output voltage are the feedback signals used in this control system. First, primary current is sensed using current transformer with turns ratio (100:1) and then sampled by a resistor to get a primary current signal (VIp); load voltage (V_{load}) signal is also sensed. To overcome the instability in current waveforms at high duty cycles, value slope compensation technique with the ramp signal of 200 kHz is added to the VIp signal to generate the primary current (Iprim); output DC voltage signal is compared with the appropriate reference value and passed to the voltage PI controller to generate the primary current reference value (Iprim-ref), then the primary current (Iprim) is compared with reference value (Iprim-ref) to generate the duty cycle of the PWM generator for the primary and secondary switches QA, QB, QC, QD, QA1, and QB1.

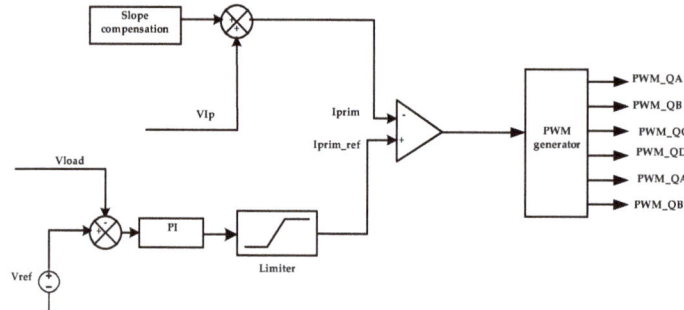

Figure 8. Technique of the phase shift PWM DC–DC converter controller.

Table 2 shows the input/output specifications and the results of the design converter main components of the DC–DC converter.

Table 2. DC–DC Converter Specifications and Design Results.

Parameter	Value	Unit
Input voltage	400 (320–410)	V
Output voltage	54 (45–63)	V
Output voltage transient (V_{tran})	1	Vpp
Rated power	2	kW
Switching frequency	100	kHz
Designed efficiency	95%	-
Input voltage hold up time (t_{hold})	4	ms
Input capacitor (C_{in})	350	uF
Transformer turns ratio	20/4/4	Np/Ns/Nt
Transformer reactance (L_{mag})	2.8	mH
Primary switches	IPW60R070CFD7	-
Secondary switches	IPP110N20N3	-
Output capacitor (C_{load})	2750	uF
Output Inductor (L_{load})	22	uH
Resonant inductor (L_r)	30	uH

3. Simulation Results and Discussions

Phase shift DC–DC converter with power density about 2 kW has been implemented in PSIM software using the designed components in the previous sections to show the system performance under different loading conditions and to test the reliability of inrush controlling circuit to limit the high inrush current at initial power application to the primary side of the power supply circuit.

With full load condition and DC input voltage about 400 V, Figure 9 shows the steady state simulation result of the DC output power curve of the DC–DC converter.

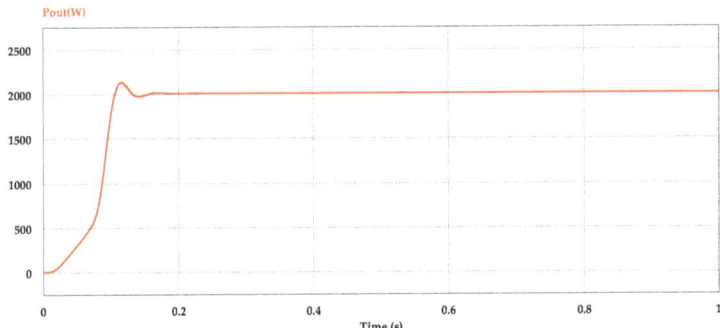

Figure 9. Converter output power curve at full load condition.

The designed power supply efficiency curve at loading condition from 10% to 100% of the rated output power (2 kW) with input voltage 400 V and switching frequency about 100 kHz is illustrated in Figure 10, which shows that the designed power supply offers full-load efficiency at about 95.1%, half-load efficiency at about 95.3%, and the maximum being three-quarter-load efficiency at about 95.6%.

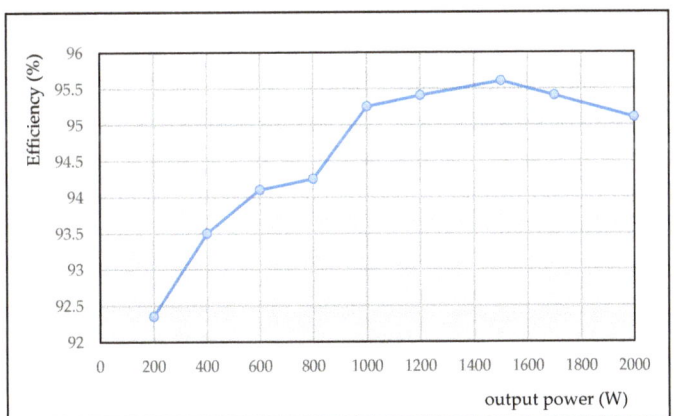

Figure 10. Converter efficiency curve with different loading condition at V_{in} = 400 V.

As mentioned before in Section 2.2, the gate charge transfer characteristics of the MOSFET switch can be used to control the slew rate of the input capacitance charging in order to control and limit the inrush current to the required value. Figure 11 shows the bulk input capacitance charging voltage curves with, and without, using the proposed inrush current control circuit, where the black curve refers to the supply DC input voltage, the red curve is the voltage curve of the input capacitance without using the proposed control circuit, and the blue curve is the voltage curve of the input capacitance using the proposed control circuit. From these curves, it is clearly observed that controlling the slew rate of VGS voltage by controlling the input capacitance of the MOSFET, leads to the reduction of the high dv/dt of the input capacitance, which subsequently reduces the inrush current at initial power application.

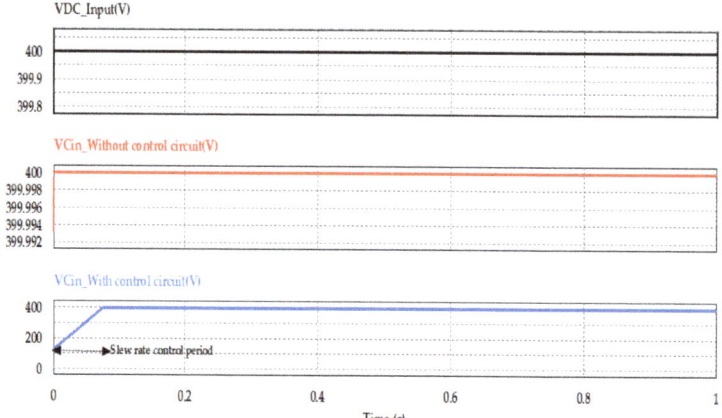

Figure 11. Bulk input capacitance charging voltage with and without using the proposed control.

Waveforms of the average value of the converter input current at switching frequency of 100 kHz with, and without, using the inrush current control circuit are shown in Figures 12 and 13, respectively. From both cases, and as the result of controlling the voltage slew rate at the initial power application to the converter circuit based on the transition time, which is required to meet the required inrush current, we can notice that the proposed controlling circuit reduced the peak inrush current from 16.10 A to about 6.40 A, which is in the allowable range (given in Table 1) of the input DC current to the primary side of DC–DC converter; moreover, it is clear to observe that the peak value of the input current at steady state with full load condition is about 8.20 A in both cases.

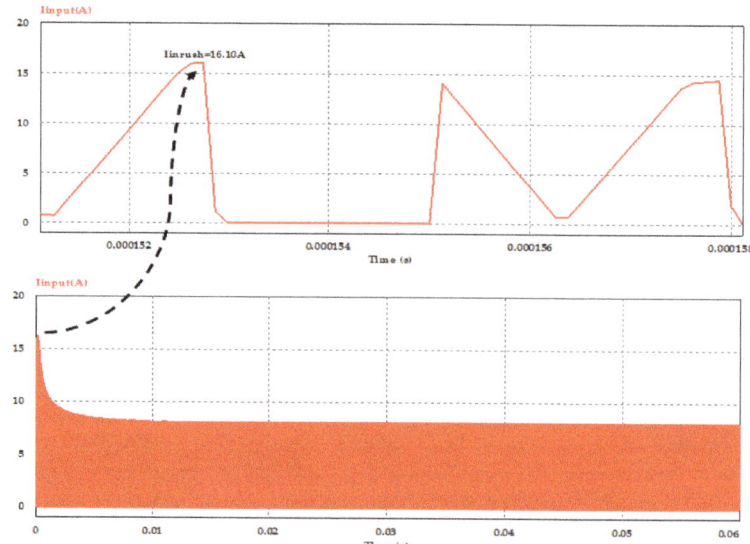

Figure 12. Converter input current without inrush current mitigation circuit at full load and V_{in} = 400 V.

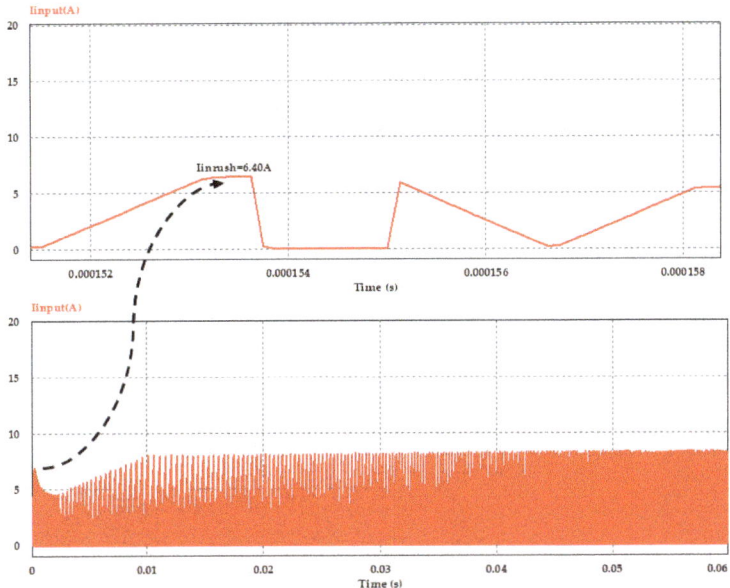

Figure 13. Converter input current with inrush current control circuit at full load and $V_{in} = 400$ V.

Figure 14 shows the VGS and Idrain characteristics of the MOSFET in the inrush current control circuit. When the VGS voltage reached to Vth (about 3.75 V) of the used MOSFET switch, the drain current starts to increase, at the point when VGS reaches V_{plt}, and circuit inrush current tries to increase but based on the gate charge transfer characteristic depicted in Figure 5, drain current saturated at constant value about 6.40 A as shown in the simulation result. Additionally, from this figure it is clearly observed that the protection of the MOSFET switch, which occurred at Region 4 when the Zener diode circuit broke down before the voltage reached 30 V (BV_{GS} of the used MOSFET) and the voltage was saturated at about 28.5 V.

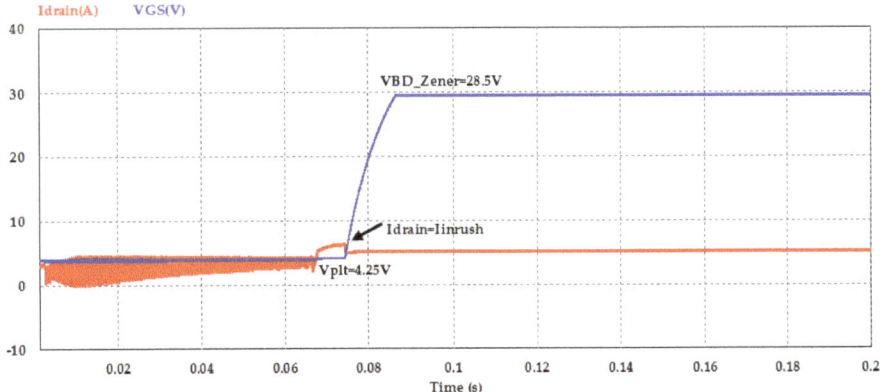

Figure 14. Gate charge waveform of the MOSFET in inrush current mitigation circuit.

Phase shift PWM controlling circuit and the resonant inductor are designed to offer ZVS at the two legs of the bridge at the primary side of the DC–DC converter switches, as shown in current and voltage waveforms of the switches (QA and QC) in Figures 15 and 16.

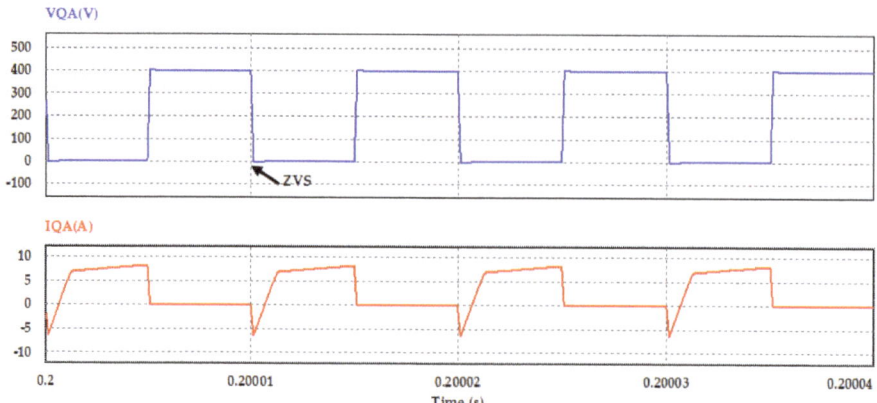

Figure 15. Voltage and current waveforms of switch QA converter primary side at $Po = 2$ kW.

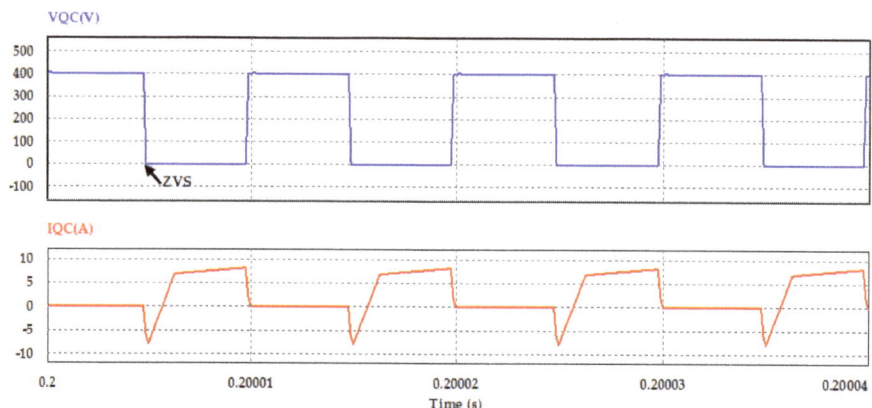

Figure 16. Voltage and current waveforms of switch QC in converter primary side at $Po = 2$ kW.

4. Experimental Setup

The complete designed converter circuit and the proposed inrush current mitigation circuit are experimentally setup and tested with the available maximum DC electronic load in the lab (1000 W), UCC28950 phase shift PWM controller IC from Texas Instruments, Dallas, Texas, United States, 2016 is used to control the DC–DC converter switches, UCC28950 IC provide 4-PWM signals with constant frequency (100 kHz) for the primary side switches, and 2-PWM for the synchronous rectification at the secondary side switches with the availability of the primary current compensation to restore the current stability and the voltage loop control to adjust the output voltage at the specified value.

The proposed inrush current mitigation circuit was tested with the practical case of the telecom two stages AC–DC power supply, where the DC–DC converter was supplied by the PFC converter stage with output voltage of 400 V DC. In order to show the reliability of the proposed inrush current control circuit in mitigation of the high current overshot in the input current to the converter, soft starter of the analog UCC28950 IC is disabled through the (SS/EN) pin and only the proposed inrush current control circuit is connected to the input side of the DC–DC converter circuit.

Figure 17 shows the appearance of the experimental setup of the complete power supply consisting of PFC and DC–DC converter stages and with using the proposed inrush current control circuit connecting in between. 1000W KIKUSI PLZ1004WH, Japan, 2019 DC electronic load was connected at load side, and 2.5 kW PFC converter stage from Infineon with average efficiency more than 95% [47], inrush current control circuit was connected between the two stages and the input AC voltage was applied to the input of the PFC stage so that the output voltage will be 400 V DC, input current to the both stages has been measured using high scale current probe FLUKE i1000s, USA, FLUKE company with scale choice of (10:1) A with, and without, connecting of the proposed inrush current mitigation circuit.

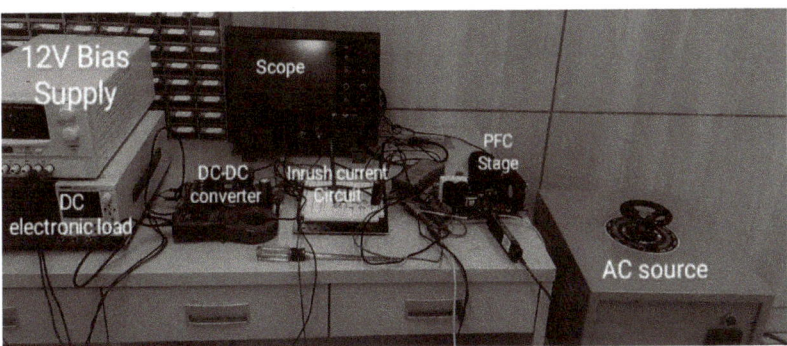

Figure 17. Experimental setup of DC–DC converter circuit with inrush current mitigation circuit.

Figures 18 and 19 show the DC–DC converter stage input current, from which we can investigated that with using inrush current control circuit, the inrush current at the startup of the converter is reduced from 20.32 A to about 4.06 A. Additionally, in both figures, differences in current shape and starting time of each waveform can be observed, this occurred due to changing in the slew rate of the input capacitance voltage, which leads to limit the inrush current at the safe limit using the inrush current controlling circuit, in Figure 19, with the current still increasing gradually until it reached to 4.06 A, at this moment, and as explained before in the controlling circuit characteristics in Figure 14, the drain current is saturated at the safe limit to prevent the inrush current from increasing to higher limits.

Figures 20 and 21 show the PFC converter stage AC input current, from which we can investigate that by connecting the proposed inrush current control circuit between the two stages of the power supply, the peak overshoot of the input current to the PFC stage will reduced from 33.55 A to 15.33 A. Additionally, the difference is noticed between starting time and waveform shapes, due to the controlling of the slew rate of the input voltage by using the proposed inrush current control circuit.

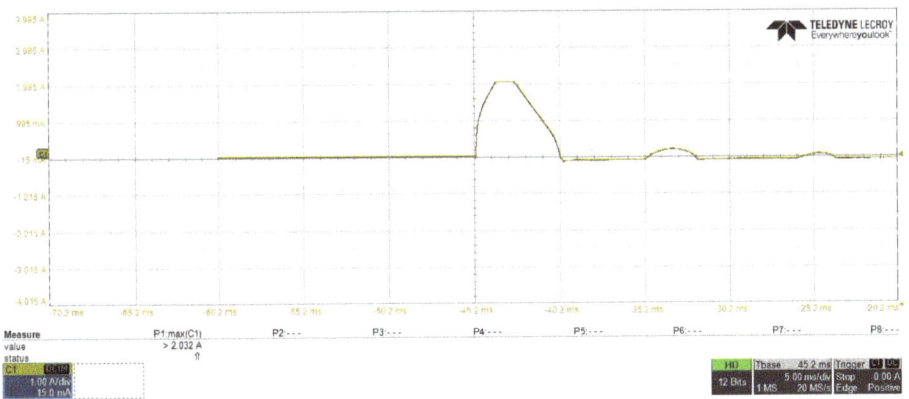

Figure 18. DC–DC converter stage input current without using of the proposed inrush current mitigation circuit.

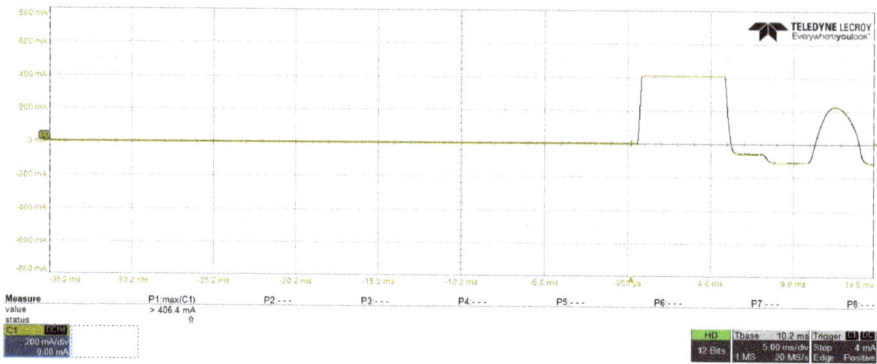

Figure 19. DC–DC converter stage input current with using of the proposed inrush current mitigation circuit.

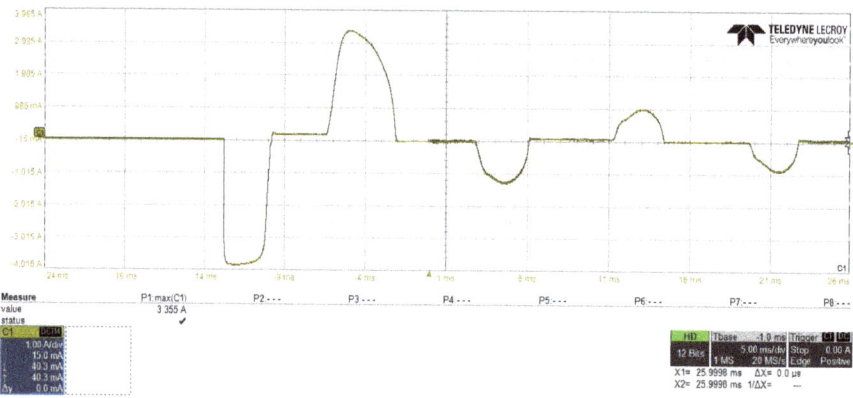

Figure 20. Power factor correction (PFC) converter stage AC input current without using of the proposed inrush current mitigation circuit.

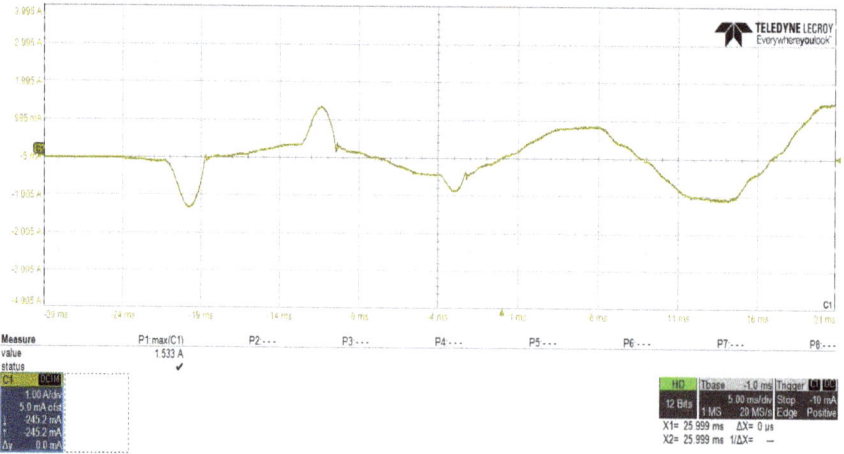

Figure 21. PFC converter stage AC input current with using of the proposed inrush current mitigation circuit.

Figure 22 shows the converter output voltage and current waveforms at steady state operation using the inrush current control circuit, in which the converter DC output power was about 1085.25 W, where the output voltage was constant at about 54.02 V and current value was about 20.090 A.

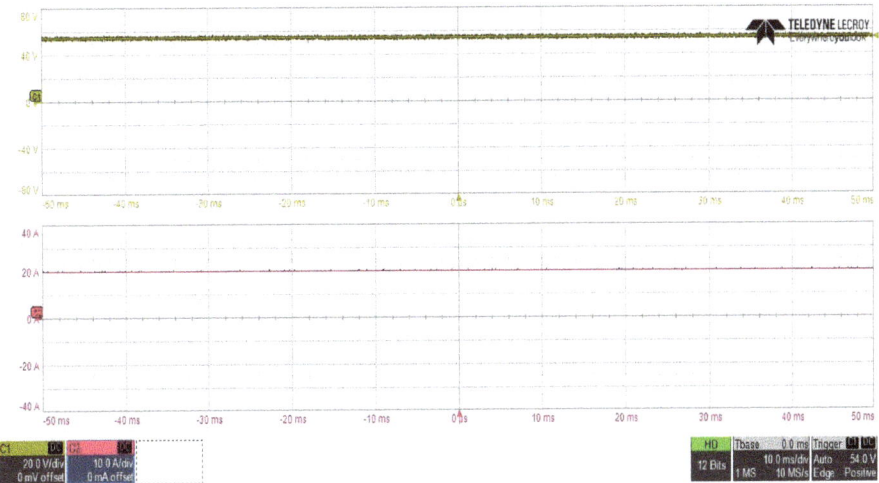

Figure 22. DC–DC converter stage output voltage and current waveforms at $V_{in} = 400$ V.

DC–DC converter output filter capacitor was designed to maintain the output voltage ripple to the specified value as explained previously in Section 3; in order to clearly observe the ripple component in the output voltage waveform, the vertical axis scale of the voltage was changed and the measurement set in scope was adjusted to measure the ripple peak to peak value and the mean value of the voltage waveform, as shown in Figure 23, which clearly shows that the peak to peak ripple voltage is about 34.4 mV (less than the designed value 200 mV) and the mean output voltage value is about 54.094 V.

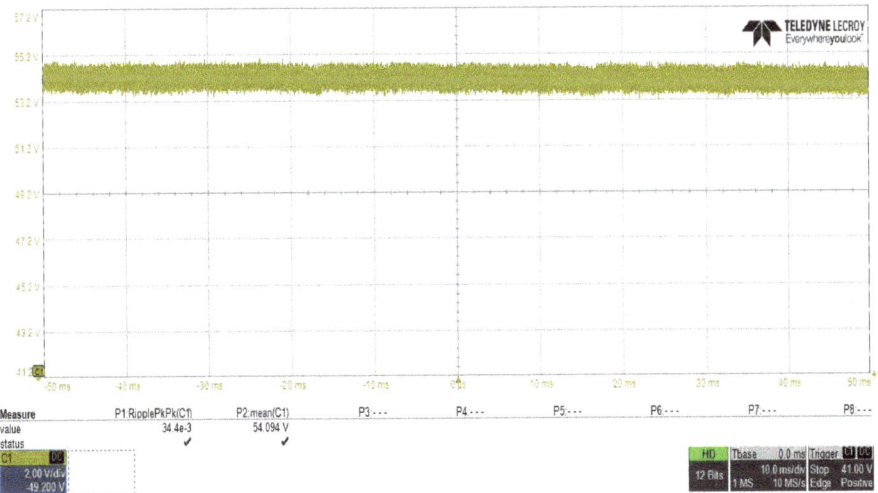

Figure 23. DC–DC converter stage output voltage ripple peak to peak measurement at $V_{in} = 400$ V.

In case of designing the DC–DC converter stage for the telecom applications, voltage controller must be implemented in order to achieve different loading condition with good voltage regulation.

Figure 24 depicts the output voltage versus output power, which shows that the designed DC–DC converter with the proposed inrush current controlling circuit also provides different loading conditions with good voltage regulation, where the voltage drop is less than 1 V (1.85%) for changing loading conditions from 200 W to 1000 W, which is less than the IEC61000-3-3 standard limit for the limitation of the voltage changes (3.3%) [48].

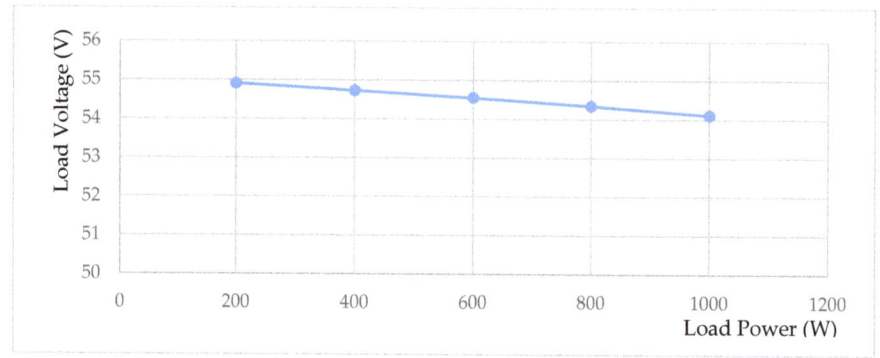

Figure 24. DC–DC converter load voltage-power characteristics at V_{in} = 400 V.

The inrush current control circuit, designed and connected between the two stages of the power supply, to control the inrush current at the initial power application to the power supply, must not effect the performance of the PFC stage at steady state operation. To study the effect of the proposed inrush current control circuit on the PFC converter performance, power analysis of the input side of the PFC converter have been performed with, and without, using the inrush current control circuit with the same loading and input voltage conditions as shown in Figures 25 and 26, which clearly shows that the reduction in the inrush current of the input supply at initial power application leads to a reduction in the supply total reactive power from 103 VA to about 90 VA, and it can also clearly be observed that the power factor value of the power supply was not affected by the connection of the proposed circuit.

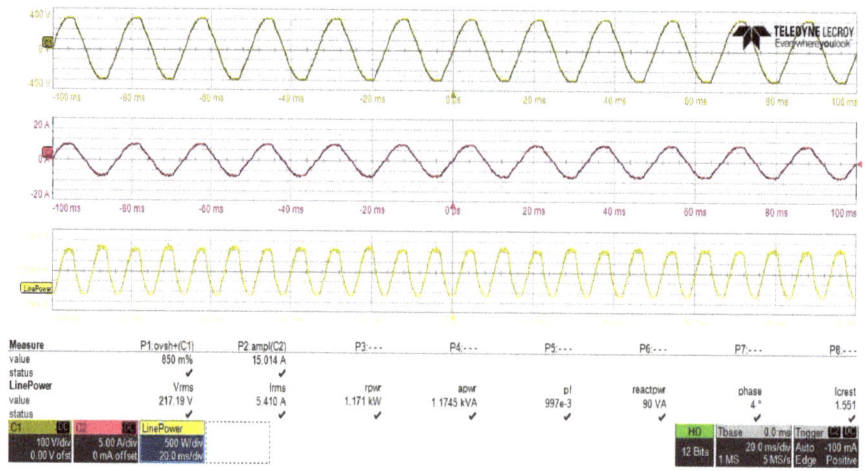

Figure 25. PFC converter stage power analysis with using of the proposed inrush current mitigation circuit.

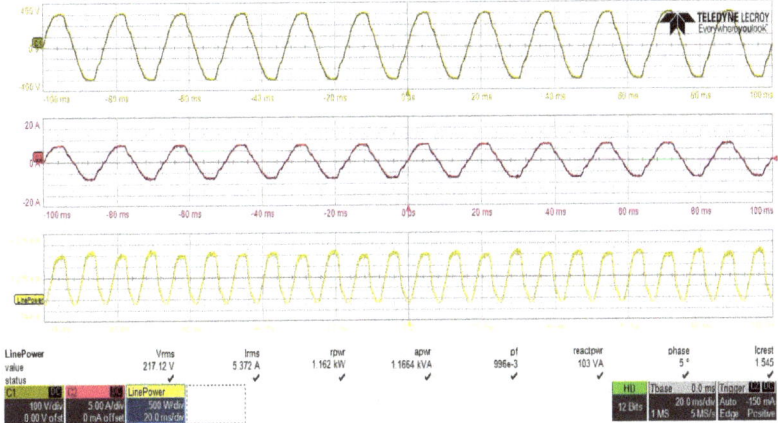

Figure 26. PFC converter stage power analysis without using of the proposed inrush current mitigation circuit.

As shown in Figure 22 the total input power to the two stages with inrush current control circuit is about 1171 W, and as shown in Figure 25, the load power is about 1085.25 W, so the power supply includes about 85.75 W power losses, distributed as 39.5 W in the PFC stage with efficiency of about 96.62% and 46.25 W in DC–DC converter and inrush current control circuit, with efficiency of about 95.73%, which is very close to the simulation results (95.40%) at the same loading. Total losses distribution in the system different parts are performed and the result is depicted in Figure 27.

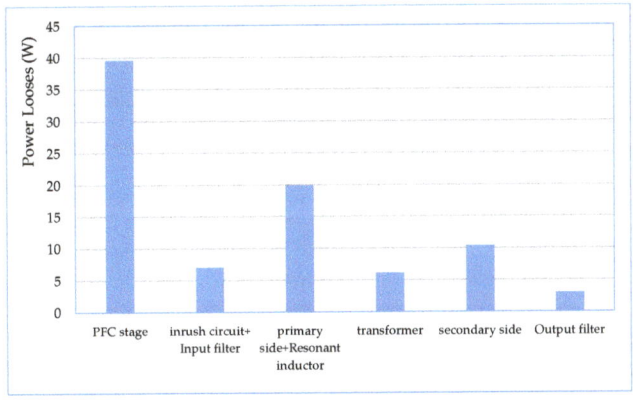

Figure 27. Losses distribution in different parts of the two stages power supply with inrush current control.

From the loss distribution of the power supply, it is noticed that the highest budget of the DC–DC converter losses was accounted by the primary side bridge and the resonant inductor with about 20 W; this loss amount is due to primary side switches internal resistance and forward voltage drop, and also due to the internal DC resistance of the resonant inductor. The second highest budget of the power losses accounted in the secondary side synchronous switches with about 10.5 W. Power loss of about 6 W was dissipated in the converter transformer primary and secondary DC resistances. Internal DC resistance of the output filter inductor and the equivalent series resistance (ESR) of the output electrolytic capacitor introduce power loss about 3 W. The remaining converter power losses budget was accounted by the proposed inrush current control circuit and the ESR of the bulk input electrolytic capacitor. One disadvantage of the proposed inrush current, is that it has an amount of about 5 W of the total supply power losses, but on the other hand, it has many advantages, which can be summarized in the following:

1. Compact design and easy implementation with cheap passive elements and single MOSFET.
2. Controllability and reliability of the circuit to limit the inrush current at the required value.
3. Reduce the current stress on the input side of the two stages of power supply at initial power application.
4. Improve the PFC stage performance by reducing the total reactive power consumption of the circuit.
5. Easy to disconnect and reconnect again between two stages of a faulty case as compared with the IC soft starter (in case where the IC soft starter function is damaged).
6. Ability to use with analog and digital control converter.

5. Conclusions

Complete design of a high power density efficient DC–DC converter stage of the telecom power supply has been discussed in this paper; a high inrush current, which is generated between two stages of power supply where the bulk input capacitor is controlled using the proposed simple inrush current control circuit designed by using single MOSFET switch and some passive elements. Simulation analysis as well as experimental setup of the practical two stages telecom power supply and inrush current control circuit is performed. Phase shifted PWM with ZVS technique is applied in DC–DC converter stage, which provided system experimentally overall efficiency about 95.73% at 50% loading condition. Furthermore, reliability of the converter designed components to achieve the performance of the telecom power supply was established; ability of the proposed inrush current control circuit to control the slew rate of the input capacitance voltage in order to reduce the high dv/dt and to mitigate the inrush current of the input current of the two stages of power supply was also achieved, which ensures a safe and smooth startup operation of the power supply.

Author Contributions: The literature review and manuscript preparation, as well as the simulations, were carried out by A.H.O. Experimental results and implementation of the prototype were carried by A.H.O. and J.B. Final review of manuscript corrections was done by N.K. and J.B. All authors have read and agreed to the published version of the manuscript.

Funding: This research was funded by grant (2018R1D1A3B0704376413) from National Research Foundation of Korea (NRF) and grant (18RTRP-B146050-01) from Railroad Technology Research Program (RTRP) funded by Ministry of Land, Infrastructure and Transport of Korean government.

Conflicts of Interest: The authors declare no conflict of interest.

References

1. Garcia, O.; Cobos, J.A.; Prieto, R.; Alou, P.; Uceda, J. Single phase power factor correction: A survey. *IEEE Trans. Power Electron.* **2003**, *18*, 749–755. [CrossRef]
2. Geddam, D.P.; Prakash, D.; Bapu, S.R.; Kumar, Y.R.; Rao, P.S.; Naidu, G.K.M.; Ramanarayanan, V. Design of 1.4 kw telecom rectifier delivering full power from 90Vac to 300Vac. In Proceedings of the IEEE INTELEC 07–29th International Telecommunications Energy Conference, Rome, Italy, 30 September–4 October 2007; pp. 670–676.
3. Kasper, M.; Bortis, D.; Deboy, G.; Kolar, J.W. Design of a Highly Efficient (97.7%) and Very Compact (2.2 kW/dm^3) Isolated AC–DC Telecom Power Supply Module Based on the Multicell ISOP Converter Approach. *IEEE Trans. Power Electron.* **2016**, *32*, 7750–7769. [CrossRef]
4. Sekar, A.; Raghavan, D. Implementation of single phase soft switched PFC converter for plug-in-hybrid electric vehicles. *Energies* **2015**, *8*, 13096–13111. [CrossRef]
5. Kim, Y.-S.; Sung, W.; Lee, B. Comparative performance analysis of high density and efficiency PFC topologies. *IEEE Trans. Power Electron.* **2003**, *6*, 2666–2679. [CrossRef]

6. Biela, J.; Badstuebner, U.; Kolar, J.W. Impact of power density maximization on efficiency of DC–DC converter systems. *IEEE Trans. Power Electron.* **2009**, *24*, 288–300. [CrossRef]
7. Yang, B.; Duarte, J.L.; Li, W.; Yin, K.; He, X.; Deng, Y. Phase-shifted full bridge converter featuring ZVS over the full load range. In Proceedings of the IECON 36th Annual Conference on IEEE Industrial Electronics Society, Glendale, AZ, USA, 7–10 November 2010; pp. 644–649.
8. Lo, Y.-K.; Lin, C.-H.; Hsieh, M.-I.; Lin, C.-H. Phase-shifted full-bridge series-resonant DC–DC converters for wide load variations. *IEEE Trans. Ind. Electron.* **2010**, *58*, 2572–2575. [CrossRef]
9. Schmidt, J.C.M.; Maragaño, J.C.; Sartori, H.C.; Pinheiro, J.R. Design methodology to achieve high-efficiency isolated ZVS DC/DC converters. In Proceedings of the 2017 IEEE 8th International Symposium on Power Electronics for Distributed Generation Systems (PEDG), Florianopolis, Brazil, 17–20 April 2017; pp. 1–8.
10. Badstuebner, U.; Biela, J.; Kolar, W.J. Design of an 99%-efficient, 5kW, phase-shift PWM DC–DC converter for telecom applications. In Proceedings of the 2010 Twenty-Fifth Annual IEEE Applied Power Electronics Conference and Exposition (APEC), Palm Springs, CA, USA, 21–25 February 2010; pp. 773–780.
11. Bai, H.; Nie, Z.; Mi, C.C. Experimental comparison of traditional phase-shift, dual-phase-shift, and model-based control of isolated bidirectional DC–DC converters. *IEEE Trans. Power Electron.* **2009**, *25*, 1444–1449. [CrossRef]
12. Xiangli, K.; Li, S.; Smedley, K.M. Decoupled PWM plus phase-shift control for a dual-half-bridge bidirectional DC–DC converter. *IEEE Trans. Power Electron.* **2017**, *33*, 7203–7213. [CrossRef]
13. Hurley, W.G.; Wölfle, W.H. *Transformers and Inductors for Power Electronics. Theory, Design and Applications*; Wiley: London, UK, 2013.
14. Koo, G.B.; Moon, G.W.; Youn, M.J. New zero-voltage-switching phase-shift full-bridge converter with low conduction losses. *IEEE Trans. Power Electron.* **2005**, *52*, 228–235. [CrossRef]
15. Kathiresan, R.; Das, P.; Reindl, T.; Panda, S.K. A novel ZVS DC–DC full-bridge converter with hold-up time operation. *IEEE Trans. Ind. Electron.* **2017**, *64*, 4491–4500. [CrossRef]
16. Das, P.; Laan, B.; Mousavi, S.A.; Moschopoulos, G. A non-isolated bidirectional ZVS-PWM active clamped DC–DC converter. *IEEE Trans. Power Electron.* **2009**, *24*, 553–558. [CrossRef]
17. Bojoi, R.; Fusillo, F.; Raciti, A.; Musumeci, S.; Scrimizzi, F.; Rizzo, S. Full-Bridge DC–DC Power Converter for Telecom applications with Advanced Trench Gate MOSFETs. In Proceedings of the 2018 IEEE International Telecommunications Energy Conference (INTELEC), Lingnotto Fiere, Italy, 7–11 October 2018; pp. 1–7.
18. Zhao, Z.; Xu, Q.; Dai, Y.; Yin, H. Analysis, Design, and Implementation of Improved LLC Resonant Transformer for Efficiency Enhancement. *Energies* **2018**, *11*, 3288. [CrossRef]
19. Tao, H.; Zhang, G.; Zheng, Z.; Du, C. Design of Digital Control System for DC/DC Converter of On-Board Charger. *J. Adv. Transp.* **2019**, *2019*, 2467307. [CrossRef]
20. Cho, J.-H.; Seong, H.-Y.; Jung, S.-H.; Park, J.-Y.; Moon, G.-U.; Youn, M.-Y. Implementation of digitally controlled phase shift full bridge converter for server power supply. In Proceedings of the 2010 IEEE Energy Conversion Congress and Exposition, Atlanta, GA, USA, 12–16 September 2010; pp. 802–809.
21. Lee, Y.J.; Bak, Y.; Lee, K.Y. Control Method for Phase-Shift Full-Bridge Center-Tapped Converters Using a Hybrid Fuzzy Sliding Mode Controller. *Electronics* **2019**, *8*, 705. [CrossRef]
22. Cho, I.-H.; Cho, K.-M.; Kim, J.-W.; Moon, G.-W. A new phase-shifted full-bridge converter with maximum duty operation for server power system. *IEEE Trans. Power Electron.* **2011**, *26*, 3491–3500. [CrossRef]
23. Moisseev, S.; Soshin, K.; Sato, S.; Gamage, L.; Nakaoka, M. Novel soft-commutation DC–DC power converter with high-frequency transformer secondary side phase-shifted PWM active rectifier. *IEE Electr. Power Appl.* **2004**, *151*, 260–267. [CrossRef]
24. Kim, Y.-D.; Cho, K.-M.; Kim, D.-Y.; Moon, G.-W. Wide-range ZVS phase-shift full-bridge converter with reduced conduction loss caused by circulating current. *IEEE Trans. Power Electron.* **2012**, *28*, 3308–3316. [CrossRef]
25. Inrush Current in DC–DC Converters. Application Notes, vpt Power. Available online: http://www.vptpower.com/wp-content (accessed on 14 November 2019).
26. Martínez-Salamero, L.; García, G.; Orellana, M.; Lahore, C.; Estibals, B. Start-up control and voltage regulation in a boost converter under sliding-mode operation. *IEEE Trans. Ind. Electron.* **2012**, *60*, 4637–4649. [CrossRef]
27. Aroudi, A.E.; Martínez-Treviño, B.A.; Vidal-Idiarte, E.; Martínez-Salamero, L. Analysis of Start-Up Response in a Digitally Controlled Boost Converter with Constant Power Load and Mitigation of Inrush Current Problems. *IEEE Trans. Circuits Syst. I* **2019**, *67*, 1276–1285. [CrossRef]

28. Mitter, S. *Active Inrush Current Limiting Using MOSFETs*; Motorola Inc.: Phoenix, AZ, USA, 1995; Available online: http://application-notes.digchip.com/010/10-13155.pdf (accessed on 10 January 2020).
29. Kalenteridis; Vasileios; Agorastou, Z.; Siskos, S. A soft start-up technique for inrush current limitation in DC–DC converters. In Proceedings of the 2019 IEEE Panhellenic Conference on Electronics & Telecommunications (PACET), Volos, Greece, 8–9 November 2019; pp. 1–4.
30. Ma, F.-F.; Chen, W.-Z.; Wu, J.-C. A monolithic current-mode buck converter with advanced control and protection circuits. *IEEE Trans. Power Electron.* **2007**, *22*, 1836–1846. [CrossRef]
31. Sathishkumar, P.; Krishna, T.N.V.; Khan, M.A.; Zeb, K.; Kim, H. Digital soft start implementation for minimizing start up transients in high power DAB-IBDC converter. *Energies* **2018**, *11*, 956. [CrossRef]
32. Managing Inrush Current. Application Report, Texas Instruments. Available online: http://www.ti.com/lit/an/slva670a/slva670a.pdf?&ts=1589053420769 (accessed on 12 January 2020).
33. Jiang, T.; Cairoli, P.; Rodrigues, R.; Du, Y. Inrush current limiting for solid state devices using NTC resistor. In Proceedings of the IEEE Southeast Conference, Charlotte, NC, USA, 30 March–2 April 2017; pp. 1–7.
34. NTC Thermistors Ordering Code. RS Components. Available online: https://docs.rs-online.com/ee83/0900766b80e7f61b.pdf (accessed on 12 January 2020).
35. Dias, V.A.; Pomilio, J.A.; Finco, S. A current limiting switch for applications in space power systems. In Proceedings of the 2017 IEEE Southern Power Electronics Conference (SPEC), Puerto Valas, Chile, 4–7 December 2017; pp. 1–6.
36. Fuengwarodsakul, N.H. Battery management system with active inrush current control for Li-ion battery in light electric vehicles. *Electr. Eng.* **2016**, *98*, 17–27. [CrossRef]
37. Marroqui, D.; Garrigos, A.; Blanes, J.M.; Gutierrez, R. Photovoltaic-Driven SiC MOSFET Circuit Breaker with Latching and Current Limiting Capability. *Energies* **2019**, *12*, 4585. [CrossRef]
38. Lee, E.-J.; Ahn, J.-H.; Shin, S.-M.; Lee, B.-K. Comparative analysis of active inrush current limiter for high-voltage DC power supply system. In Proceedings of the 2012 IEEE Vehicle Power and Propulsion Conference, Seoul, Korea, 9–12 October 2012; pp. 1256–1260.
39. Musumeci, S. Gate charge control of high-voltage Silicon-Carbide (SiC) MOSFET in power converter applications. In Proceedings of the 2015 International Conference on Clean Electrical Power (ICCEP), Taormina, Italy, 16–18 June 2015; pp. 709–715.
40. Manolarou, M.; Kostakis, G.; Manias, S.N. Inrush current limiting technique for low-voltage synchronous DC/DC converters. *IEE Electr. Power Appl.* **2005**, *152*, 1179–1183. [CrossRef]
41. Manousaka, E. DC–DC Buck Converter with Inrush Current Limiter. Master's Thesis, Faculty of Applied Sciences, Delft University of Technology, Delft, Wales, October 2013.
42. UCC28950 600-W, Phase-Shifted, Full-Bridge Application Report. Texas Instruments. Available online: http://www.ti.com/lit/slua560 (accessed on 8 March 2020).
43. N-CHANNEL 500 V –0.105Ω –31A TO-247 Zener-Protected MOSFET. Available online: https://www.st.com/resource/en/datasheet/cd00045267.pdf (accessed on 5 April 2020).
44. "Power Electronics Handbook" Book; Available online: http://site.iugaza.edu.ps/malramlawi/files/RASHID_Power_Electronics_Handbook.pdf (accessed on 25 February 2020).
45. Phase-Shifted Full-Bridge, Zero-Voltage Transition Design Considerations. Application Report. Available online: https://www.ti.com/lit/pdf/slua107 (accessed on 8 March 2020).
46. Instruments Texas. *Modeling, Analysis and Compensation of the Current-Mode Converter*; Notes, Application; Texas Instruments Inc.: Dallas, TX, USA, 1999.
47. 2.5kW PFC Evaluation Board with CCM PFC Controller ICE3PCS01G. Design Report, Infineon Company. Available online: https://www.infineon.com/cms/en/product/evaluation-boards/eval_2k5w_ccm_4p_v3/ (accessed on 15 April 2020).
48. IEC/EN61000 Standards for Power Supplies. SL Power Application Note. Available online: https://slpower.com/App-slpower/images/whitepapers/AN-G007_EN61000.pdf (accessed on 12 August 2020).

© 2020 by the authors. Licensee MDPI, Basel, Switzerland. This article is an open access article distributed under the terms and conditions of the Creative Commons Attribution (CC BY) license (http://creativecommons.org/licenses/by/4.0/).

Article

Methods of Modulation for Current-Source Single-Phase Isolated Matrix Converter in a Grid-Connected Battery Application

Goh Teck Chiang * and Takahide Sugiyama

Toyota Central R&D Labs Inc., Nagakute City 480-1192, Japan; t-sugiyama@mosk.tytlabs.co.jp
* Correspondence: tcgoh@mosk.tytlabs.co.jp

Received: 29 May 2020; Accepted: 21 July 2020; Published: 27 July 2020

Abstract: This paper discusses three methods of modulation for a single-phase isolated matrix converter. The matrix converter is combined with a transformer integration to perform power decoupling control in order to reduce the number of component and capacitor volumes. Due to the reason of (i) Alternating current (AC/AC) direct conversion and (ii) transformer integration, obtaining a clean sinusoidal grid current waveform in the modulation of matrix converter (MC) is important. Three methods of modulation are compared in terms of control complexity, quality waveform, and inductive-capacitive-inductive (LCL) filter sizing. The principal control of each method is described. Finally, a prototype was tested to verify the validity and the effectiveness of grid current control and power decoupling in the spoken circuit structure.

Keywords: AC/AC conversion; decoupling control; modulation

1. Introduction

The rapidly expanding growth of battery storage system (BSS) has urged high demands for a single-phase power converter. Figure 1a shows the applications such as home energy management system (HEMS), uninterruptible power supply (UPS), and small-scale datacenter, which uses a single-phase power converter as an interface between BSS and grid (AC 80–240 V 50/60 Hz). These applications require isolation and typically rate from 1–3 kW with a high voltage battery (100–300 V). As the price of battery is expected to reach a new low in the near future, a low cost and small size single-phase power converter is highly demanded.

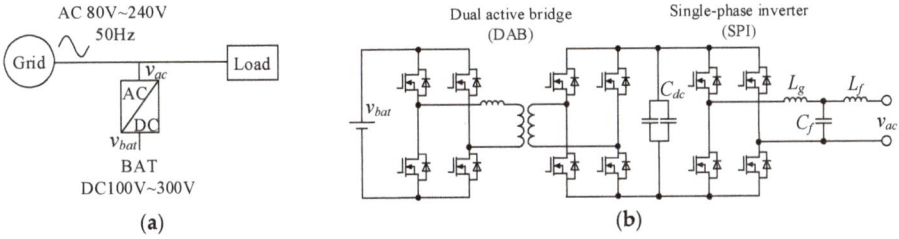

Figure 1. (**a**) Single-phase power converter for a grid-connected battery application. (**b**) Conventional circuit structure consists of a dual-active bridge (DAB) and a single-phase inverter (SPI).

Figure 1b shows the conventional circuit for a single-phase power converter. The circuit is composed of a dual-active bridge (DAB), a single-phase inverter (SPI), and an LCL filter [1,2]. The size reduction of a single-phase power converter is challenging because of using many passive components.

Several studies have focused on reducing the size of inductive component such as inductor and transformer by using a high frequency technique [3–5]. However, the major size of the converter is occupied by the capacitors C_{dc} that are used to absorb the single-phase power fluctuation. Due to the reason of current limitation in the electrolytic capacitor, capacitors are connected in parallel to form a big capacitor bank in order to absorb the single-phase power fluctuation.

The matrix converter (MC) shows a promising solution for size reduction because the capacitor can be removed [6,7]. Hence, MC can convert high frequency transformer voltage (i.e., 50 kHz) to low frequency voltage (i.e., 50 Hz), at the same time controlling the current flow bidirectional. However, for a single-phase application, a low frequency current that contains twice of the grid frequency occurrs in the battery side due to the direct AC/AC conversion. Depending on the type of battery, such as lithium battery, the single-phase fluctuation in the battery needs to be eliminated in order to protect the battery from overvoltage.

Single-phase active power decoupling techniques have been discussed and reported for compensating the single-phase power fluctuation [8,9]. A power decoupling circuit can be considered to add between the battery and full-bridge inverter (FBI) in order to compensate the single-phase low frequency current. Figure 2 shows a conventional circuit structure that consists of a power decoupling circuit, a FBI, and a MC. The power decoupling circuit consists of an inductor L_b, a capacitor C_b, a diode, and two switching devices. The single-phase power fluctuation is compensated by charging and discharging the capacitor C_b according to the grid phase angle. As a result, the single-phase power fluctuation can be eliminated with a smaller capacitor than the conventional capacitor bank.

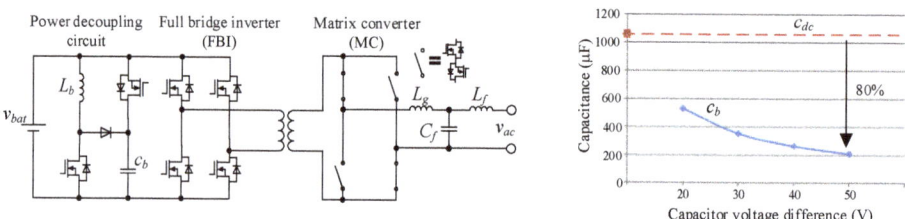

Figure 2. Circuit structure consists of a power decoupling circuit, a full-bridge inverter (FBI), and a matrix converter (MC).

The required capacitance in a single-phase converter can be defined by Equation (1).

$$C_b \geq \frac{P_{cap}}{2 \times \pi \times 2 \times f_g \times \Delta v_{cb} \times v_{cb}}; \Delta v_{cb} = v_{cbmax} - v_{cbmin} \quad (1)$$

where v_{cb} is the average battery voltage, Δv_{cb} is the capacitor voltage difference, and f_g is grid frequency. Figure 2 shows that a 1 kW calculation (capacitor power (P_{cap}) is half of the rated power), by using the power decoupling control to increase the capacitor voltage difference Δv_{cb} to 50 V, the required capacitance can be reduced by 80% comparing that to the conventional circuit at the same rated power.

However, the major drawback is that this circuit requires extra switching devices and passive component. Here, an integration technique that utilized the center-tapped of a transformer has been discussed, as shown in Figure 3 [10,11]. The center-tapped of the transformer is utilized by connecting the passive components (L_b and C_b) in order to perform the power decoupling control. Then, the FBI controls the high frequency transformer voltage and the capacitor voltage at the same time, therefore the switching devices in the power decoupling circuit can be reduced.

Without the transformer integration, the power decoupling circuit can be individually controlled and the method of modulation for the matrix converter is rather simple. Literature reviews [12–14] have demonstrated several valid modulations for the MC, where a good quality waveform can be obtained without the need for concern for the power decoupling. However, when the transformer integration is applied with MC, the modulation for MC has to be changed in order to synchronize with the FBI to obtain a proper voltage period. The failure of obtaining a clean sinusoidal waveform can distort the battery current inherently, because this capacitor cannot absorb the current fluctuation.

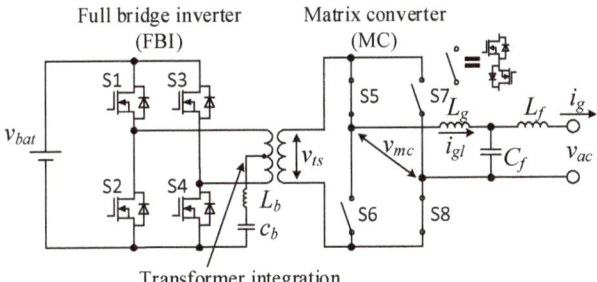

Figure 3. Circuit structure consists of a full-bridge inverter (FBI) with a transformer integration decoupling control and a matrix converter (MC).

This paper discusses three methods of modulation for the MC that is applied with the transformer integration. The first conventional method is a carrier comparison with a D-FF and the second conventional method is a delta-sigma conversion based on pulse density modulation (PDM) which have been addressed in [15–17]. This paper introduces a third method which is a carrier comparison with a zero-vector commutation, and further discusses the difference of each method. The details of each method is described individually. Then, the comparisons among these methods in term of (i) control complexity, (ii) waveform quality, and (iii) LCL sizing are discussed. Then, the validity of the modulation along with comparison results is shown. Finally, a 1 kW prototype was tested to show the validity of the power decoupling with a MC.

2. Control Scheme

2.1. System Control

The system control block diagram is shown in Figure 4. The control is divided into two parts: (i) Voltage and current closed-loop controls in FBI and (ii) grid voltage and current with a phase locked loop (PLL) in MC. The FBI is performed as a voltage source to control the high frequency transformer voltage v_{ts} and capacitor voltage v_{cb} at the same time. Then, the high frequency transformer voltage is fed into the MC, and therefore MC is performed as a current source converter to control the grid current.

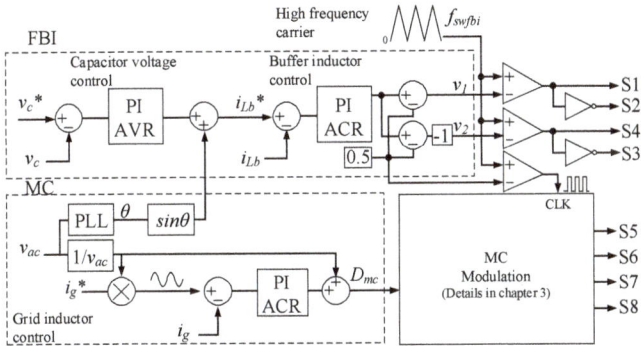

Figure 4. System control block diagram, full-bridge inverter (FBI) controls the voltage level and matrix converter (MC) controls the current for the single-phase power converter.

In FBI, a low time response of automatic voltage regulator (AVR) is applied to control the capacitor voltage proportionally to half of the battery voltage. Then, the grid phase angle which is calculated from the PLL in the grid control is added to control the phase angle of the inductor current. Then, a high time response current control (automatic current regulator ACR) is applied into the inductor current control. Note that the input of phase angle in the FBI control is also used to enable or disable the power decoupling control (where 0 is disabled).

Since the MC is a current source controller, a normalized grid voltage command is feed-forwarded into the controller. Then, a high time response current control is applied into the grid inductor current, where a sinusoidal duty command (D_{mc}) is used to generate the corresponding gate signals.

Figure 5 shows the principal control of the power decoupling. The relationship among the grid power p_g, battery power p_{bat}, and capacitor power p_{cb} is defined in Equation (2), where p_{avg} is the average power and μ_o is the grid phase angle. When subjected to the frequency of the grid power, the battery current contains twice the grid frequency.

$$p_g = p_{bat} - p_{cb}; \; p_{cb} = p_{avg} \cos(2\omega_o t) \qquad (2)$$

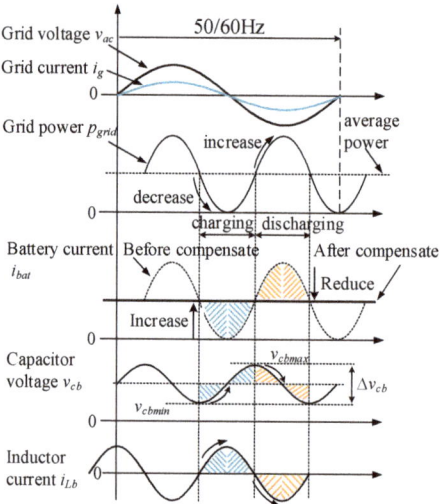

Figure 5. Principle control of power decoupling, charging and discharging states in the capacitor are used to compensate the single-phase current that occurs in the battery current.

In the power decoupling, the capacitor power is divided into a charging and discharging state. When the grid power is lower than the average power, this period is known as a charging state. During the charging state, the battery power is loaded into the capacitor by controlling the center-tapped inductor current. Then, the capacitor voltage difference Δv_{cb} increases from v_{cbmin} to v_{cbmax} during this period.

When the grid power is higher than the average power, this period is known as a discharging state. During the discharging state, the previously charged power in the capacitor is discharged by the center-tapped inductor current. Then, the capacitor voltage difference Δv_{cb} decreases from v_{cbmax} to v_{cbmin} during this period. Since no power delivery is needed from the battery, the battery current remains at its average value. By repeating these two cycles according to the grid phase angle, the capacitor voltage difference is controlled to compensate the single-phase current in the battery.

2.2. Switching Behaviors in FBI

The switching behavior and current relationships in FBI are described. Figure 6 shows the relationships between the two current components (DC and AC) in the FBI. The DC current component occurs when the DC/DC conversion is performed between the battery voltage and capacitor voltage. In this case, the FBI is equivalent to a buck converter with a 180 degree phase shift. Two DC currents, i_{Lb1} and i_{Lb2}, are manipulated with the duty to control the center-tap connected inductor current and capacitor voltage, which can be defined in Equations (3) and (4).

$$v_{cb} = D_{fbi} \times v_{bat}; \quad (3)$$

$$i_{Lb} = i_{Lb1} + i_{Lb2}; \quad (4)$$

where v_{cb} is the capacitor voltage, v_{bat} is the battery voltage, D is the duty of FBI, and i_{Lb} is the inductor current.

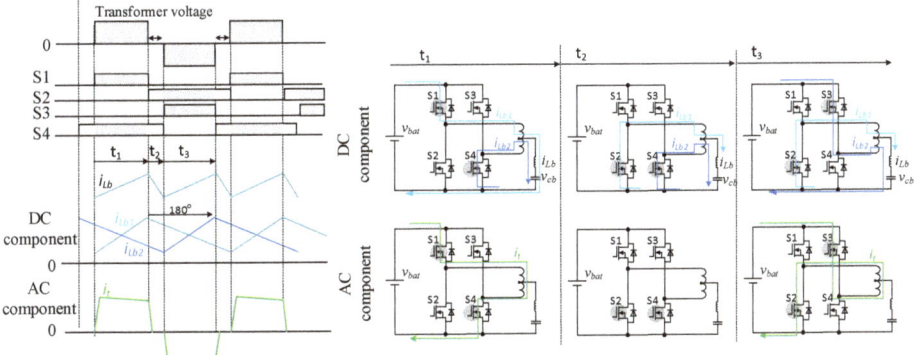

Figure 6. Relationship between the two current components in FBI.

On the other hand, the AC current component occurs when the battery power delivers the grid via the transformer, which is the transformer current. The AC current component is controlled with corresponding to the modulation of MC, which is equivalent to the grid inductor current. Therefore, the relationship between the transformer current and grid inductor current can be expressed as Equation (5).

$$i_{tp} \times N = i_{ts} = i_{gl}; \quad (5)$$

where i_{tp} is the primary side current (FBI), i_{ts} is the secondary side current (MC), N is the transformer ratio, and i_{gl} is the grid inductor current.

According to the state of the capacitor (charging or discharging) and the amplitude of the battery current, the total of four switching behaviors can be summarized as shown in Figure 7. The zero-voltage periods of FBI (S1S3 or S2S4 are turned on) are utilized to discharge and charge the inductor current.

During the discharging state, the battery current needs to be reduced and therefore the charged energy in the capacitor C_b discharges to the battery side. When S2S4 are turned on, the current circulates via S2 and S4 to keep the charged energy. When S1S3 are turned on, the charged energy in the inductor is released to the battery via S1 and S3. The current cancellation between the battery current and inductor current reduces the high peak of the battery current.

During the charging state, the battery current needs to be increased and therefore the low peak of the battery current charges into the capacitor C_b. Here, when S1S3 are turned on, the battery current flows via S1 and S3 to charge inductor Lb. Then, when S2S4 are turned on, the charged energy in the inductor circulates via switching devices.

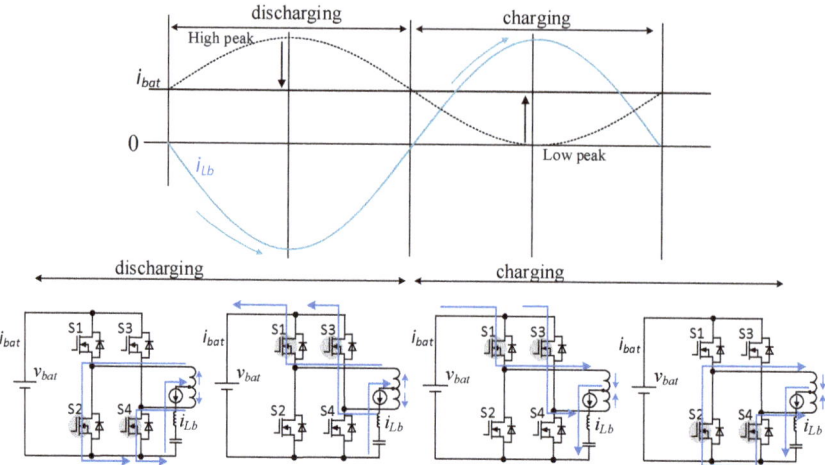

Figure 7. Switching behavior of full-bridge inverter (FBI) according to the state of capacitor and amplitude of the battery current.

2.3. Modulation in MC

In MC, a low frequency voltage pulse width is formed to control the grid inductor by accumulating from the high frequency transformer voltage v_{ts}. The high frequency transformer is controlled by FBI which is magnetized from the battery voltage. As shown in Figure 8, the switching sequence is divided into positive and negative voltage periods according to the polarity of the grid voltage. Then, each of these voltage periods is implemented with zero-vector periods in order to discharge the grid inductor current. As a result, the method of modulation is used to control the length of these voltage periods in order to control the 50 Hz grid inductor current sinusoidal.

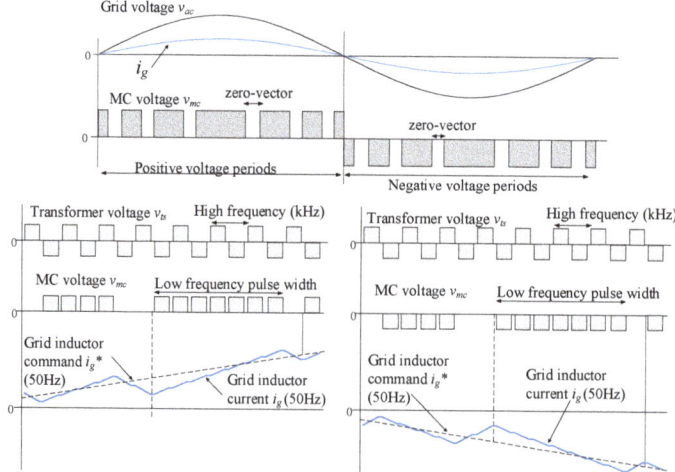

Figure 8. Modulation in matrix converter (MC), zero-vector periods are implemented in both positive and negative voltage periods.

Figure 9a shows the switching behavior in MC, which can differ to normal switching states and zero-vector periods. When the gate signals S5S8 are turned on, the grid inductor is induced by the positive transformer voltage to charge the grid inductor. Otherwise, when the gate signals S6S7 are turned on, the grid inductor is then induced by the negative transformer voltage. On the other hand, when the gate signals S5S7 (or S6S8) are turned on, known as the zero-vector periods, a circulating loop is created inside the switching devices to allow the grid inductor current to circulate and discharge the energy in the grid inductor.

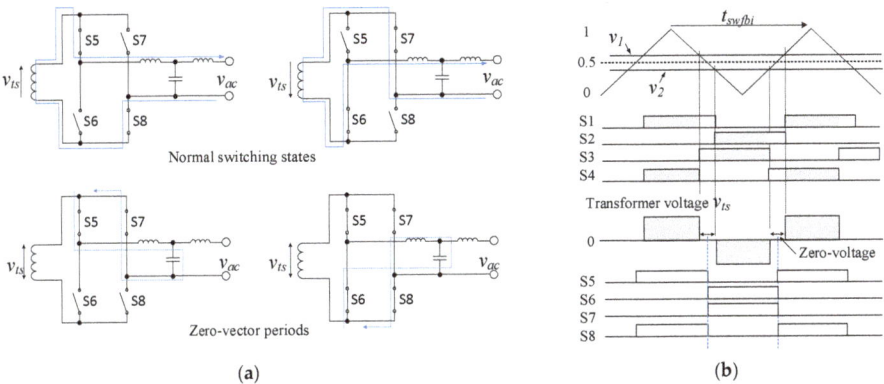

Figure 9. Switching behaviors in matrix converter (MC). (**a**) Normal switching states and zero-vector periods. (**b**) Zero-voltage switching (ZVS) relationship between full-bridge inverter (FBI) and matrix converter (MC).

Furthermore, Figure 9b illustrates the zero-voltage switching (ZVS) relationships between FBI and MC. Since the transformer voltage is magnetized from the battery voltage, a three-level high frequency transformer voltage can be produced. That is, when gate signals S1 S3 or S2 S4 in FBI are turned on, no voltage-product occurrs in the MC. These zero-voltage periods are utilized in the switching intervals of MC to reduce the switching loss. During the gate-off transition, the drain-source voltage of switching devices drops to zero before the gate signal is turned off. Then, during the gate-on transition, the gate signal is turned on before the voltage is applied to the switching devices. Therefore, both of the transitions can achieve ZVS.

However, the leakage inductance of the transformer needs to be taken into consideration during the switching intervals. The energy in the leakage inductance needs to be discharged while the transformer current changes the direction. Here, the approach is to use the grid inductor current to cancel out with the leakage inductance current during the switching intervals. Figure 10 explains and illustrates the phenomenon, where the positive transformer voltage is changed to the negative transformer voltage while the grid side produces a positive voltage.

As shown in Figure 10, the transformer voltage becomes zero before the switching intervals start. Then, following that the S6AB and S7AB are turned on in the next switching interval. During this state, the leakage inductance current is used to discharge the capacitance S5A and also charge the capacitance S6B. At the same time, the grid inductor current is flowing via S6AB in a reverse direction, as a result the leakage inductance current and grid inductor current cancel out each other. The same phenomenon applies to S7AB and S8AB, capacitance S7B is charged and capacitance S8A is discharged by the leakage inductance current. Then, the grid inductor current is flowing via S7BA in an opposite direction to achieve the current canceling.

As a result, forming an accurate voltage period, achieving ZVS, and current cancelling at the same time is important in the method of modulation.

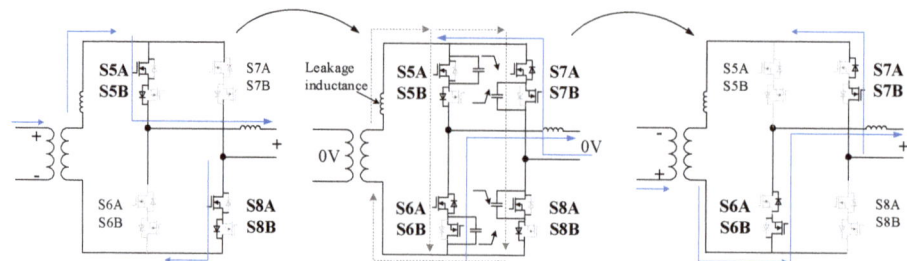

Figure 10. Current cancelling in matrix converter (MC) to discharge the leakage inductance current during switching intervals.

3. Methods of Modulation

3.1. Carrier Comparison with D-FF (D-FlipFlop)

The first method is to use a D-flipflop (D-FF) function, the control block diagram is shown in Figure 11. A carrier comparison with D_{mc} is used to generate two sets of switching signals SPQ and SNQ. When SPQ and SNQ are both turned on zero-vector periods are formed. These two switching signals are inputted to a D-FF, where the D-FF is synchronized with the CLK, and a XNOR logic is applied to produce gate signals for S5–S8. The CLK is used to synchronize the switching intervals of MC with the zero-voltage periods of FBI in order to achieve ZVS.

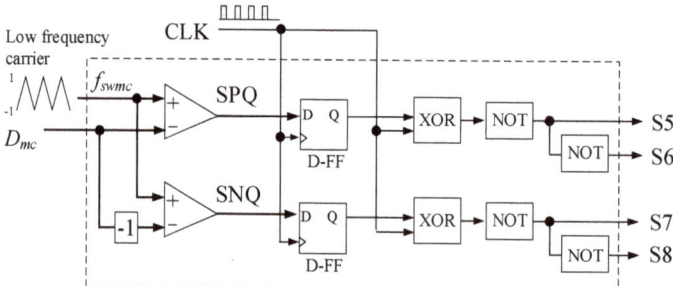

Figure 11. Carrier comparison with the D-flip flop block diagram.

However, D-FF creates a voltage error due to the occurrence of improper time length. Figure 12 shows the relationships among switching signals SPQ SNQ, gate signals S5–S8, and voltage pulse width. First, SPQ and SNQ form the required voltage pulse width accordingly based on the carrier comparison. After the SPQ and SNQ are aligned with D-FF, the voltage pulse width applied to the grid inductor either becomes longer or shorter than the original voltage pulse width. These improper pulse widths create voltage errors and the average grid inductor current is misadjusted. As a result the grid current fluctuates irregularly.

3.2. Delta-Sigma Conversion with Pulse Density Modulation (PDM)

In order to eliminate the voltage error, a delta-sigma conversion which is based on pulse density was discussed. Figure 12 shows the control block diagram and Figure 13 shows the relationship between duty D_{mc} and quantization error Q_r. The carrier comparison is not applied because the integral changes corresponding to the quantization error. One cycle of the quantization level is equivalent to one cycle of the CLK. The Q_r is obtained based on the differential value between the D_{mc} and D_{mc}. Note that the amplitude of D_{mc} does not change according to the grid current command ($i_g{}^*$) but the level of quantization error changes depending on the pulse density. As shown in Figure 14, the original middle point is $D_{mc} = 0.5$. Then, the level of Q_r changes depending on the pulse density that is used to form the grid current command, which is $Q_r > D_{mc}$ or $Q_r < D_{mc}$. The comparison between the D_{mc} and Q_r produces the corresponding voltage signals Sa and Sb in order to produce the desired voltage pulse width. EXOR logic is applied to Sa and Sb to synchronize with CLK in order to produce gate signals.

That is, when i_g needs to increase, a longer voltage pulse width is required and therefore Q_r gets higher than D_{mc}. On the other hand, when i_g needs to decrease, a shorter voltage pulse is required and Q_r gets lower than D_{mc}.

These phenomenon are illustrated in Figure 15, where (a) $Q_r < 0.5$ and (b) $Q_r > 0.5$. In Figure 15a, in order to decrease the grid current, most of the Q_r periods are lower than D_{mc}, then Sa produces a short voltage signal only when Q_r is higher than D_{mc}. On the other hand, in Figure 15b, in order to increase the grid current, a longer voltage pulse is required. Notice that the level of Q_r increases, and most of the Q_r periods are higher than D_{mc} to produce the desired voltage pulse width. As a result, the grid current can be controlled sinusoidal without the voltage error, and ZVS can be achieved by synchronizing to CLK.

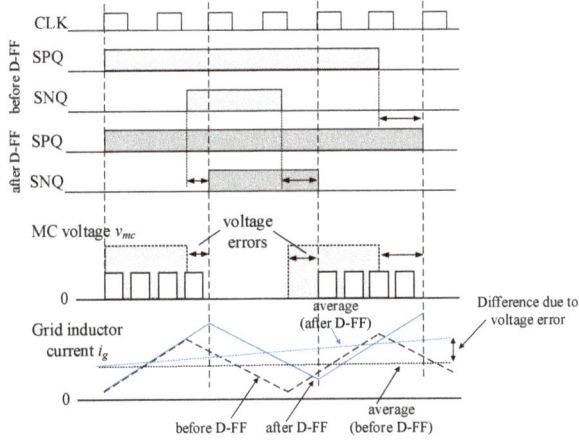

Figure 12. Relationship among CLK, SPQ, SNQ, MC voltage, and the grid inductor current are shown to demonstrate the voltage error in D-flip flop.

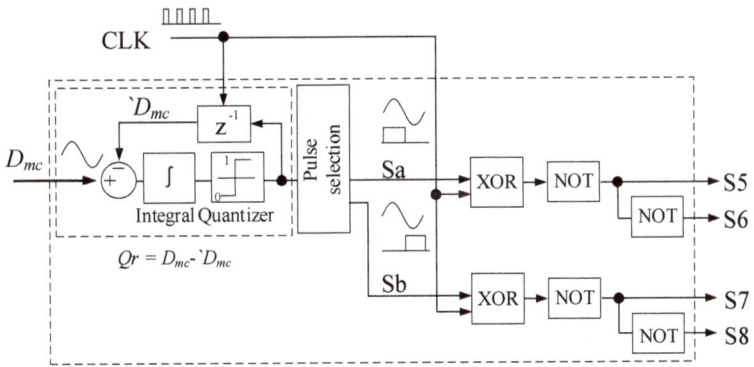

Figure 13. Delta-sigma conversion block diagram.

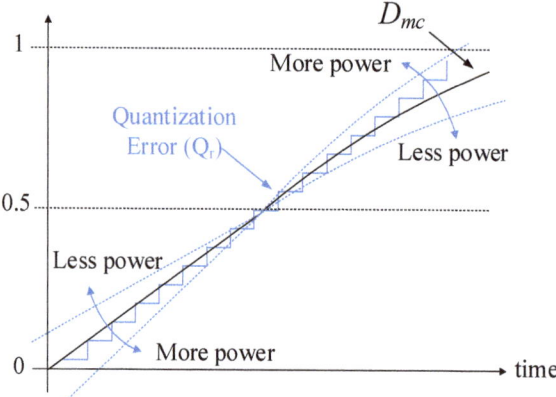

Figure 14. Principle control of delta-sigma, where quantization error changes according to the power level in order to obtain the desired pulse width ($Q_r > D_{mc}$ or $Q_r < D_{mc}$).

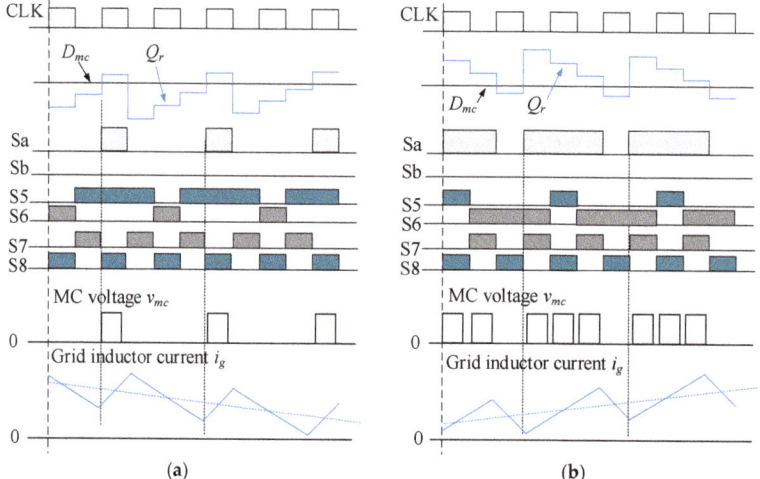

Figure 15. Relationship among CLK, D_{mc}, Q_r, MC voltage, and the grid inductor to demonstrate the control of delta-sigma. (**a**) $Q_r < 0.5$. (**b**) $Q_r > 0.5$.

However, without the carrier comparison the integral resets the value depending on the quantization level at a random frequency, as shown in Figure 16. As a result, the grid current ripple has an inconsistent frequency which causes a resonance problem during the low output power [18]. Furthermore, the resonance also occurs in the battery current due to the AC/AC direct conversion. Note that this resonance cannot be compensated in the single-phase power decoupling, therefore one approach is to decrease the cut-off frequency of the LCL; however, the size of LCL needs to increase as a drawback.

3.3. Carrier Comparison with Zero-Vector Commutation

The method of carrier comparison with zero-vector commutation is shown in Figure 17. This control is implemented with a constant frequency and a commutation to eliminate voltage error. A carrier comparison which is based on the pulse width modulation (PWM) is used and compared with D_{mc} to generate a constant frequency voltage pulse width, similar to D-FF. Then, a zero-vector determination (FS-SYN) is used to distinguish between the normal switching states and zero-vector periods. During the normal switching states, the CLK synchronizes the switching timing so that each of the switching intervals of MC can achieve ZVS.

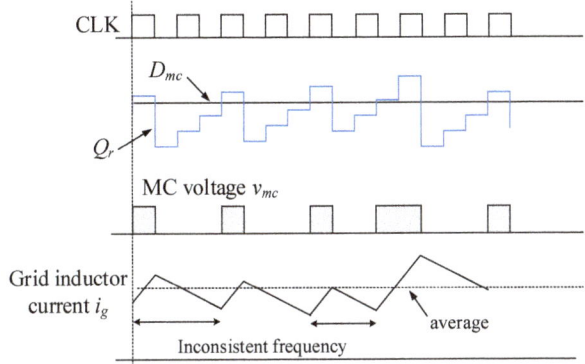

Figure 16. Inconsistent frequency in the grid inductor current due to the quantization error.

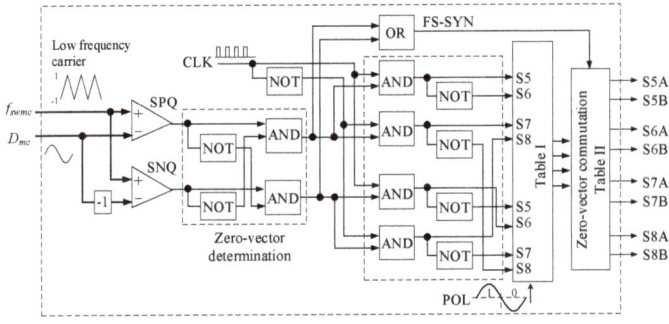

Figure 17. Carrier comparison with zero-vector commutation is introduced to overcome the voltage error and the inconsistent frequency problem.

During the zero-vector periods, since the transformer voltage is applied on the switching devices, hard-switching will cause the voltage at the switching device. In order to prevent the short-circuit state, the transformer current first needs to be blocked before switching. Furthermore, a current circulating path must first be created in order to achieve current cancelling.

The zero-vector commutation is applied only to the first and last switching intervals of the zero-vector periods. A total of six categories are divided in the zero-vector commutation which depends on the polarity of the transformer voltage, as shown in Figure 18. That is, if the zero-vector period occurs from a positive transformer voltage and ends on a positive voltage or ends on a negative voltage, it is known as PV-to-Z, PVZ-to-PV or PVZ-to-NV, respectively. On the other hand, if the zero-vector period occurs from a negative transformer voltage and ends on a positive voltage or ends on a negative voltage, it is known as NV-to-Z, NVZ-to-PV or NVZ-to-NV, respectively. Then, a two-step commutation is performed to circulate and cancel out the leakage inductance current. The switching algorithms of normal switching states are summarized in Table 1, and the switching algorithms of zero-vector commutation are summarized in Table 2.

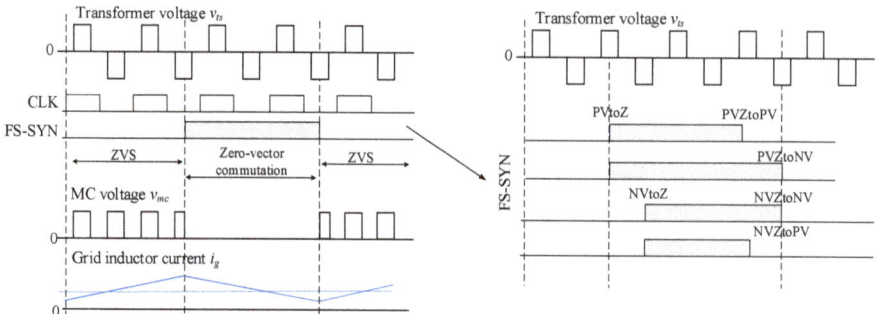

Figure 18. Categories of zero-vector commutation depending on the polarity of transformer voltage.

Table 1. Switching algorithms of normal switching states (ZVS).

Symbols	Polarity = 1			Polarity = 0		
SPQ	1	1	1	0	0	0
SNQ	0	0	1	1	1	1
Vts	+	-	+/-	+	-	+/-
S5	1	0	0	0	1	1
S6	0	1	1	1	0	0
S7	0	1	0	1	0	1
S8	1	0	1	0	1	0
Vmc	+	+	0	-	-	0

Table 2. Switching algorithms of zero-vector commutation.

Actions	PV-to-Z		PVZ-to-PV		PVZ-to-NV		NV-to-Z		NVZ-to-PV		NVZ-to-NV	
Sequence	1st	2nd	1st	2nd	1st	2nd	1st	2nd	1st	2nd	1st	2nd
S5A	1	1	1	1	1	0	0	0	1	1	0	0
S5B	1	1	1	1	0	0	0	0	0	1	0	0
S6A	0	0	0	0	1	1	1	1	0	0	1	1
S6B	0	0	0	0	0	1	1	1	1	0	1	1
S7A	0	1	0	0	1	1	0	0	0	0	1	1
S7B	1	1	1	0	1	1	1	0	0	0	0	1
S8A	1	0	0	1	0	0	1	1	1	1	1	0
S8B	0	0	1	1	0	0	0	1	1	1	0	0

Figure 19 shows the switching sequence of the zero-vector commutation for the case PV-to-Z. First, S7B is turned on and S8B is turned off, the transformer current continues to flow in the same direction via the S8A and S8B diode. Then, as S8A is turned off, the capacitance in S8A is charged by the leakage inductance current in order to build up the blocking voltage. Thus, the capacitance in S7A is discharged to reduce the blocking voltage. At the same time, the grid inductor current flows in the opposite direction via S7AB and therefore both currents are cancelled out with each other in S7AB. After S8 is completely turned off the grid inductor current starts circulating via S5AB and S7AB, the zero-vector period is created in the grid side.

Figure 20 shows the switching sequence of the zero-vector commutation for the case PVZ-to-NV. First, S6A is turned on and S5B is turned off, the circulating current continues to flow in the same direction via S5AB and S7AB. Then, as S6B is turned on, the capacitance in S5B is charged by the transformer current to build up the blocking voltage and capacitance in S6B is discharged. At the same time, the grid inductor current is cancelled out with the transformer current in S5AB. Since S5B is completely turned on, S5A is turned off at no loss and the transformer current starts to flow to the grid via S6AB and S7AB.

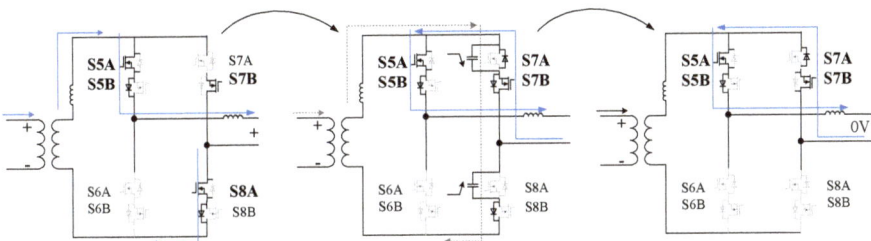

Figure 19. Switching sequence for PV-to-Z, current cancelling in S7AB.

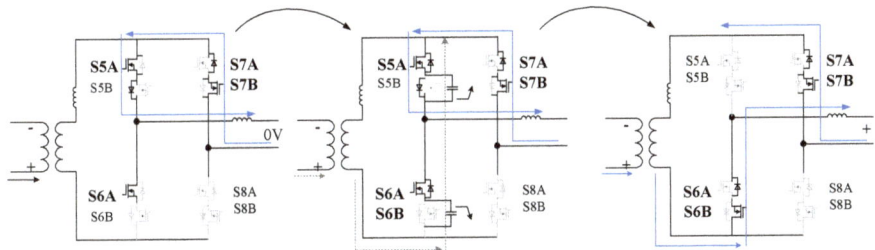

Figure 20. Switching sequence for PVZ-to-NZ, current cancelling in S5AB.

On one hand, for the case of PVZ-to-PV, current cancelling cannot be performed due to the polarity of transformer voltage. The differential current is creating a voltage surge but a short-circuit state is not created and therefore a breakdown of the device does not happen. Figure 21 explains the phenomenon of the switching sequence. First, S7A is turned off and S8B is turned on. Next, the capacitance in S8A needs to discharge in order to allow the current flows. Therefore, as S7B is turned off, the transformer current is charging the capacitance in S7A at the same time discharging the capacitance in S8A. Note that the only circulating path for the grid inductor current is via S5AB and S8AB and therefore the grid inductor current flows with the transformer current via S8AB in the same direction. As a result, the voltage surge occurs while turning on S8AB.

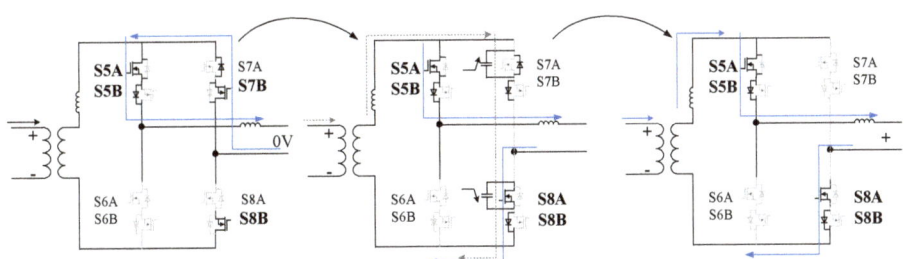

Figure 21. Switching sequence for PVZ-to-PV, current cancelling is not achieved.

Figures 22–24 illustrate the switching sequence for the case of negative transformer voltage. The principle control of current cancelling is similar, Figure 22 shows the switching sequence for NV-to-Z. While the transformer current is flowing via S6AB and S7AB, S8A is first turned on. Then, as S7B is turned off, the capacitance in S7B is charged and the capacitance S8B is discharged by the transformer current. At the same time, the grid inductor current flows via S8AB and current cancelling can be achieved in S8AB during this state.

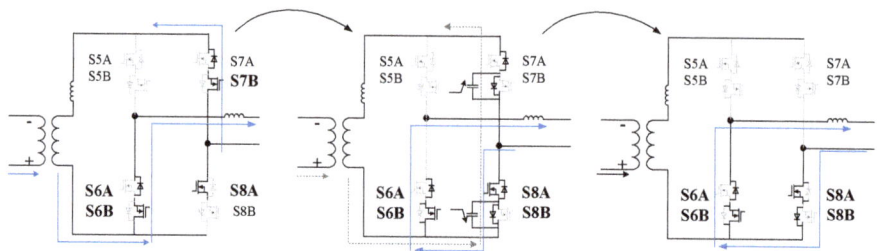

Figure 22. Switching sequence of NV-to-Z, current cancelling in S8AB.

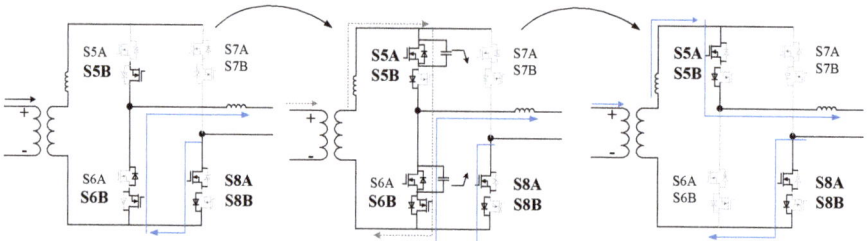

Figure 23. Switching sequence of NVZ-to-PV, current cancelling in S6AB.

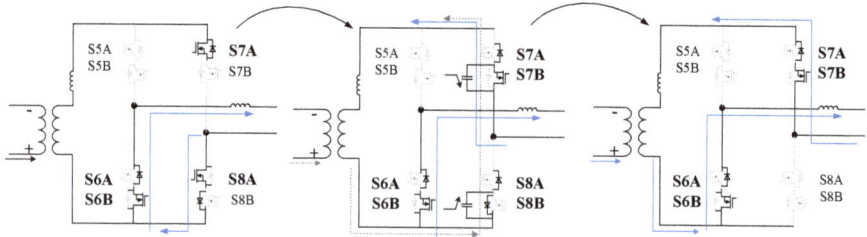

Figure 24. Switching sequence of NVZ-to-NZ, current cancelling is not achieved.

Figure 23 shows the switching sequence for NVZ-to-PV. While the grid inductor current is circulating via S6AB and S8AB, S5B is turned on and S6A is turned off. Then, as S5A is turned on, the capacitance in S5A is discharged and the capacitance in S6A is discharged by the transformer current. The grid inductor current flows in an opposite direction in S6AB to achieve current cancelling. Since the S6A is completely turned on, the transformer current starts to flow via S5AB and S8AB. On the other hand, similar to PVZ-to-PV, current cancelling cannot be achieved in NVZ-to-NV. As shown in Figure 24, as S8B is turned on, the capacitance in S8B is charged and the capacitance in S7B is discharged by the transformer current. Due to this reason, the grid inductor current flows in the same direction with the transformer current in S7AB, and the voltage surge occurs during this interval.

Note that this zero-vector commutation differs from the traditional commutation in MC [19,20]. The traditional commutation is applied to form the voltage pulse width, however the zero-vector commutation is applied during the zero-vector periods only (no voltage-product). Therefore, the voltage error that occurred in the traditional commutation is not a concern. The purpose of the zero-vector commutation is to cancel the current while charging and discharging the capacitance in the switching devices, which is simpler than the traditional commutation.

4. Simulation Results

The comparisons among the methods of modulation are demonstrated in the simulation results. The simulation parameters of each method are summarized in Table 3, which is similar to the experimental parameters. Moreover, the proportional-integral (PI) gain control for each method has been tuned to provide the best result. Figure 25 shows the relationships between transformer voltage, MC voltage, and grid inductor current based on the three modulations: (a) Carrier comparison with D-FF, (b) delta-sigma conversion with PDM, and (c) carrier comparison with zero-vector commutation, respectively.

Table 3. Simulation/experimental parameters.

Names	Symbol	Value
Battery voltage	v_{bat}	100–200 V
Grid voltage	v_{ac}	100 V 50 Hz
Transformer voltage ratio	$N_{fbi}:N_{mc}$	1:1.75
FBI switching frequency	fsw_fbi	100 kHz
Capacitor	C_b	400 µF
Inductor	L_b	10 µH
MC switching frequency	fsw_mc	10 kHz
Filter inductor	L_f	50 µH
Filter capacitor	C_f	22 µF
Grid inductor	L_g	425 µH

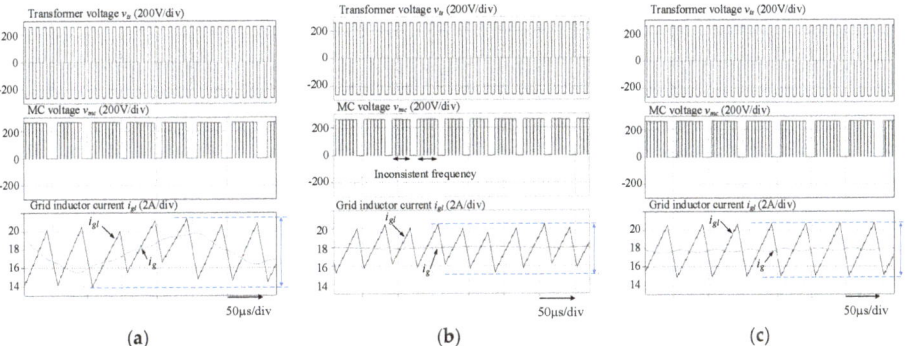

Figure 25. Simulation results that demonstrate the relationship between the matrix converter (MC) voltage pulse width and grid inductor current with different modulation methods. (a) D-flip flop, (b) delta-sigma conversion, (c) zero-vector commutation.

In the D-FF, it can be noticed that due to the misalignment of pulse width, the average of the grid current cannot be constantly controlled. As a result, the fluctuation of grid inductor current is the largest among the three methods. In the delta-sigma conversion, the voltage error can be resolved and therefore the fluctuation of grid inductor current is smaller than D-FF. However, it can be confirmed from the voltage pulse width that it has an inconsistent frequency due to the level of quantization error. With the zero-vector commutation, the voltage error can be eliminated and the fluctuation of grid inductor current is removed due to containing a consistent frequency in the current ripple. Therefore, the waveform quality can be improved as compared to the other two methods.

Figure 26 shows the simulation results at a low output power (300 W) that demonstrates the waveform of the battery current with all the three methods. Due to the direct AC/AC conversion, the distortion of grid current directly affects the waveform of the battery current. Notice that in Figure 26a, the battery current is heavily distorted in the D-FF because of the voltage error. In the case of delta-sigma conversion as shown in Figure 26b, the distortion in the battery current can be greatly reduced because the voltage error has resolved. However, a resonant frequency of the LCL filter occurrs at the battery current because of inconsistent frequency in the grid inductor current. As shown in Figure 26c, the zero-vector commutation can solve the two above problems. The distortion and resonance frequency in the battery current can both be removed due to a clean sinusoidal waveform that can be achieved in the grid inductor current.

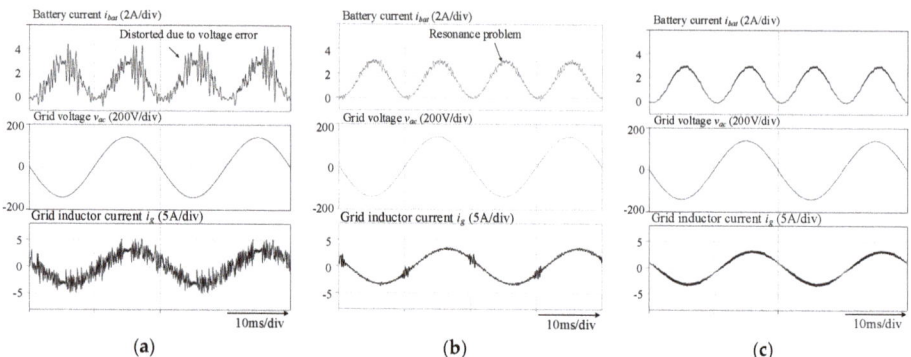

Figure 26. Low power (300 W) simulation result to demonstrate the differences of each method of modulation. (**a**) D-flip flop, (**b**) delta-sigma conversion, (**c**) zero-vector commutation.

Figure 27 shows another operating waveform at larger power (3.3 kW) to demonstrate the waveform of battery current with all the three modulation methods. D-FF is shown to have the worst distortion in battery current among the three methods. On the other hand, delta-sigma can achieve a clean sinusoidal waveform in the grid current nearly to the zero-vector commutation. This is because the peak-peak current is limited by the cut-off frequency of LCL filter, as the amplitude of the grid current becomes larger, the ripple current that is caused by the resonant frequency has lesser effect compared to the low power.

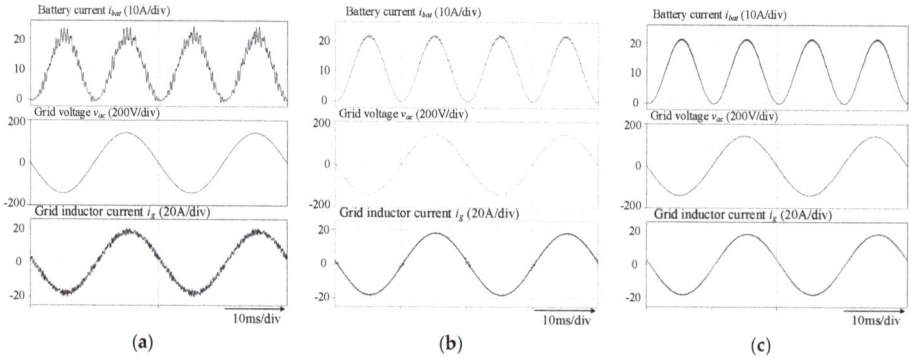

Figure 27. High power (3.3 kW) simulation result to demonstrate the difference of each method of modulation. (**a**) D-flip flop, (**b**) delta-sigma conversion, (**c**) zero-vector commutation.

Figure 28 shows the comparison of grid current THD among these modulations at low and high output power, respectively. The LCL cut-off frequency is regulated from 3 to 7.5 kHz by adjusting inductor L_f and the C_f capacitor while keeping the same impedance percentage. The results in Figure 28a (low power) shows that D-FF has the highest THD, and only the zero-vector commutation can reach the THD below 5% at a cut-off frequency of 5 kHz. In Figure 28b (high power), both the delta-sigma conversion and zero-vector commutation can reach the THD below 5% within the cut-off frequency from 3 to 7.5 kHz.

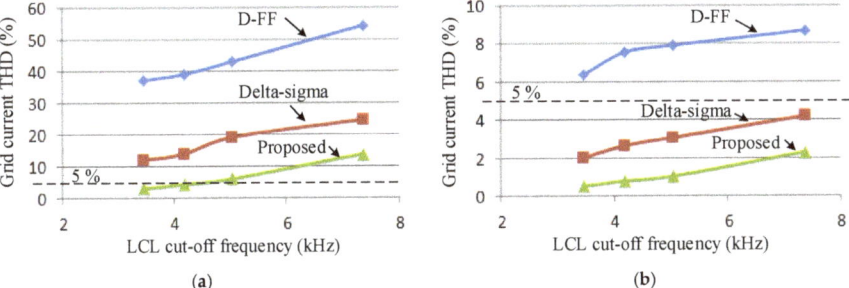

Figure 28. Comparison of grid current total harmonic distortion (THD) with the different inductive-capacitive-inductive (LCL) cut-off frequency between low power and high power. Zero-vector commutation achieves the lowest total harmonic distortion (THD) regardless of power level. (a) Low power (300 W). (b) High power (3k W).

As a result, in order for the delta-sigma control to achieve 5% THD at low output power, the cut-off frequency needs to be reduced with the penalty of a larger size in the LCL filter. Therefore, the zero-vector commutation can achieve the smallest size in the LCL filter and also achieve THD below 5% for both low and high output power, among the three methods.

5. Experimental Results

Figure 29 shows the layout of the 1 kW prototype (203 × 113 × 10 mm). Switching devices are placed on both sides (left: MC; right: FBI), then a planer transformer is placed in the middle and the capacitor C_b (400 µF) and an inductor L_b (10 µH) that is connected to the center point of transformer is placed on the top side of the prototype.

Figure 29. Layout of a 1 kW prototype.

Figure 30 shows the effectiveness of ZVS and zero-vector commutation. Figure 30a shows the result before applying zero-vector commutation. During the first (PV-to-Z) and last (PVZ-to-PV) switching intervals of zero-vector periods because hard-switching happens at a short-circuit state, therefore over-voltage occurs at the transformer voltage. On the other hand, during the ZVS periods it can be confirmed that voltage spikes did not occur at the transformer voltage because switching devices are aligned to the zero-voltage of the transformer to achieve ZVS.

Figure 30b shows the result after applying the zero-vector commutation. As shown in the result, the voltage spike at the transformer voltage can be greatly reduced in PV-to-Z, comparing that to Figure 30a. The zero-vector commutation enables the switching state to go into the zero-vector periods to allow the current to circulate inside a loop and achieve current cancelling. As a result, the voltage spike of the switching device during the zero-vector periods can be resolved. On one hand, current cancelling cannot be achieved in PVZ-to-PV and therefore the voltage surge occurs on the transformer voltage. Since the short-circuit state can be prevented, the voltage surge is smaller than that compared to Figure 30a.

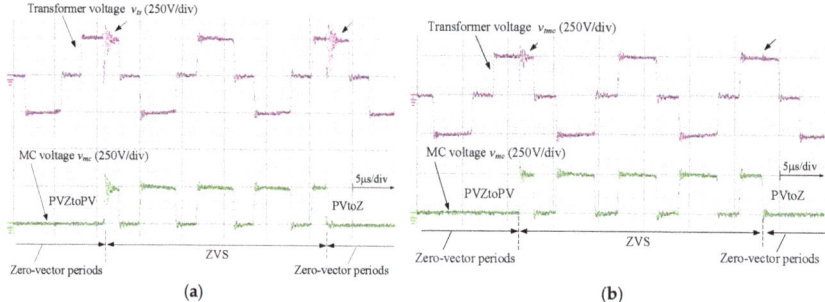

Figure 30. Experimental result to demonstrate the effectiveness of the zero-voltage switching (ZVS) and zero-commutation. (a) Without the zero-vector commutation. (b) With the zero-vector commutation.

The comparison of experimental results at low output power (300 W) between the delta-sigma and zero-commutation is shown in Figure 31a,b. In this result, the power decoupling was disabled in order to validate the effectiveness of the method of modulation. The capacitor voltage v_{cb} is controlled at 90 V constantly with a battery voltage of 180 V. In Figure 31a, we noticed that the grid inductor current i_{gl} has a huge current ripple due to the inconsistent frequency. As a result, the battery current fluctuates at a resonance frequency of 1.25 kHz. Figure 30b shows the results obtained by the zero-vector commutation, the distortion in the battery current can be nearly eliminated because a clean sinusoidal grid inductor current can be obtained.

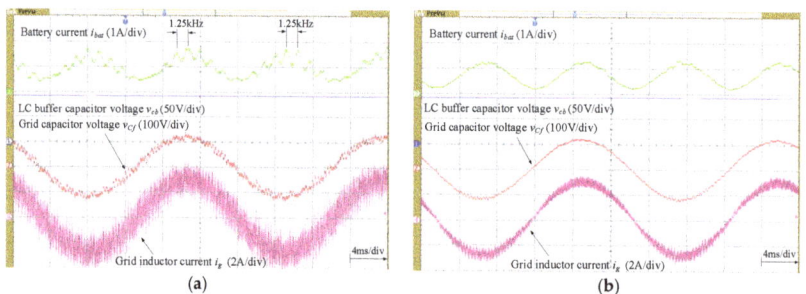

Figure 31. Comparison of experimental results between delta-sigma and zero-vector commutation. (a) Delta-sigma shows fluctuation in the battery current at 1.25 kHz. (b) Zero-vector commutation shows a clean sinusoidal waveform that can be obtained in the grid inductor current and battery current.

Figure 32a,b shows the fast Fourier transform (FFT) analysis of the grid current between the delta-sigma and the zero-vector commutation, respectively. Even the number of harmonic components contains the battery current. Then, it can be noticed that the 4th, 6th, and 8th harmonic components in the delta-sigma conversion is higher than that of the zero-vector commutation. The result can confirm that the zero-vector commutation could achieve better THD than the conventional ones.

Figure 33 shows the effectiveness of power decoupling with the zero-vector commutation. In Figure 33a, the power decoupling control was disabled and therefore the battery current contains a low frequency component. Then, after being applied to the power decoupling control as shown in

Figure 33b, the single-phase power fluctuation occurs in the capacitor voltage. The average capacitor voltage is constantly kept at half of the battery voltage, then Δv_{cb} of approximately 30 V is controlled at 100 Hz sinusoidal waveform to compensate the single-phase power fluctuation. This is also identical to the theoretical calculation which is explained in Figure 3, where a 400 µF capacitor with 30 V voltage difference is designed for the power decoupling. As a result, the single-phase power fluctuation can be reduced in the battery current. Note that the low frequency fluctuation occurrs in the battery current because of the DC bias effect in the ceramic capacitor.

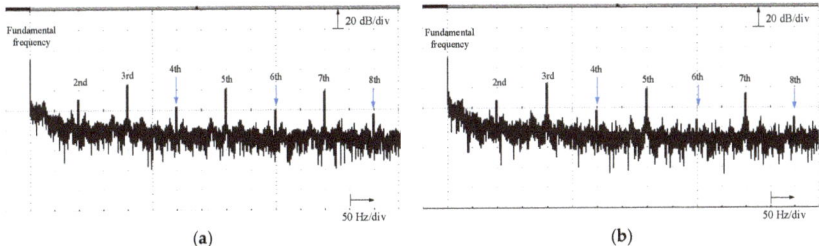

Figure 32. Comparison of the fast Fourier transform (FFT) analysis on grid current between delta-sigma and zero-vector commutation, the even harmonic component is lower in the zero-vector commutation. (a) Delta-sigma. (b) Zero-vector commutation.

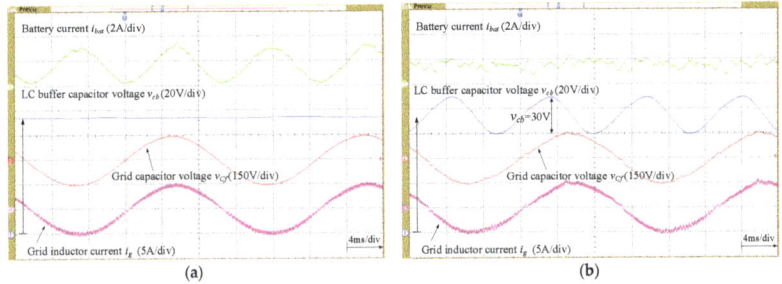

Figure 33. Experimental results to demonstrate the effectiveness of power decoupling control with the zero-vector commutation. (a) Without the decoupling control; capacitor voltage is constantly controlled at the average of battery voltage. (b) With the decoupling control; the capacitor voltage difference is 30 V with the decoupling control to compensate the single-phase power fluctuation.

Figure 34 shows the experimental measurement efficiency of the prototype. The prototype achieves the highest efficiency 91.5% at 1 kW. Optimization of the losses will be considered in the future work to improve the efficiency.

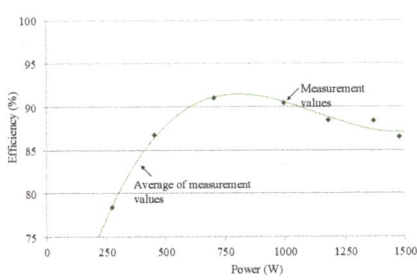

Figure 34. Measurement of efficiency, the prototype achieves the highest efficiency of 91.5%.

6. Conclusions

The comparisons among the three methods and along with other literature reviews are summarized in Table 4. Power decoupling is obviously not considered in the past studies due to the difference of circuit structure. The control also shows a difficult level due to complex commutation rules. This paper describes and compares three methods of modulation of MC for the power decoupling with the transformer integration. Zero-vector communication is introduced in this paper. The effectiveness of these modulations have been demonstrated in simulation and experimental.

Table 4. Comparison results among the three methods of modulation.

Modulation Methods	Power Decoupling	Control Complexity	Quality Waveform (Grid Current THD, Battery Current Ripple)		LCL Filter Sizing
			Low-Power	High-Power	
SPWM synchronous rec. [12]	×	△	○	○	50 kHz L_f: 10 µH C_f:1 µF
PWM four-step comm. [13]	×	×	○	○	5 kHz L_f: 300 µH C_f:4.7 µF
PWM comm. [14]	×	×	△	○	N/A
D-FF [15]	○	○	×	△	× 3 kHz L_f: 100 µH C_f:22 µF
Delta-sigma [16]	○	△	△	○	△ 2 kHz L_f: 100 µH C_f: 45 µF
Zero-vector commutation	○	△	○	○	○ 5 kHz L_f: 50 µH C_f:22 µF

○ = good, △ = average, × = poor.

Modulations that consider the power decoupling are summarized as follows. The D-FF is simple in terms of control but the quality waveform is poor due to the voltage error. Delta-sigma achieves average among the three, in order to improve the THD during low-power, a bigger size of LCL filter is required. The zero-vector commutation can produce a better quality waveform but the control complexity requires a high-bandwidth controller. If the design level is only concerned for high-power, delta-sigma with a lower bandwidth controller is another option of choice.

Author Contributions: Conceptualization, G.T.C. and T.S.; methodology, G.T.C.; software, G.T.C.; validation, G.T.C. and T.S.; formal analysis, G.T.C.; investigation, G.T.C.; resources, G.T.C.; data curation, G.T.C.; writing—original draft preparation, G.T.C.; writing—review and editing, G.T.C. and T.S.; visualization, G.T.C.; supervision, T.S.; project administration, T.S.; funding acquisition, T.S. All authors have read and agreed to the published version of the manuscript.

Funding: This research received no external funding.

Conflicts of Interest: The authors declare no conflict of interest.

References

1. Raggl, K.; Nussbaumer, T.; Doerig, G.; Biela, J.; Kolar, J.W. Comprehensive design and optimization of a high-power density single-phase PFC. *IEEE Trans. Ind. Electron.* **2009**, *56*, 2574–2587. [CrossRef]
2. Choi, W.; Rho, K.-M.; Cho, B.-H. Fundamental duty modulation of dual-active bridge converter for wide-range operation. *IEEE Trans. Power Electron.* **2016**, *31*, 4048–4606. [CrossRef]
3. Jovanovic, M.M.; Jang, Y. State-of-the art, single-phase, active power factor correction techniques for high power applications. *IEEE Trans. Ind. Electron.* **2009**, *56*, 2574–2587. [CrossRef]
4. Musavi, F.; Eberle, W.; Dunford, W.G. A high-performance single-phase bridgeless interleaved PFC converter for plug-in hybrid electric vehicle battery chargers. *IEEE Trans. Ind. Appl.* **2011**, *47*, 1833–1843. [CrossRef]

5. Xue, L.; Shen, Z.; Boroyevich, D.; Mattavelli, P. GaN-based high frequency totem-pole bridgeless PFC design with digital implementation. In Proceedings of the IEEE Applied Power Electronics Conference and Exposition (APEC), Charlotte, NC, USA, 15–19 March 2015; pp. 759–766.
6. Kolar, J.W.; Friedli, T.; Rodriguez, J.; Wheeler, P.W. Review of three-phase PWM AC-AC converter topologies. *IEEE Trans. Ind. Electron.* **2011**, *58*, 11. [CrossRef]
7. Empringham, L.; Kolar, J.W.; Rodrigues, J.; Wheeler, P.W.; Clare, J.C. Technological issues and industrial application of matrix converters: A review. *IEEE Trans. Ind. Electron.* **2013**, *60*, 10. [CrossRef]
8. Sun, Y.; Liu, Y.; Su, M.; Xiong, W.; Yang, J. Review of active power decoupling topologies in single-phase systems. *IEEE Trans. Power Electron.* **2016**, *31*, 4778–4794. [CrossRef]
9. Komeda, S.; Fujita, H. A power decoupling control method for an isolated sing-phase AC-to-DC converter based on direct AC-to-AC converter topology. *IEEE Trans. Power Electron.* **2018**, *33*, 9691–9698. [CrossRef]
10. Itoh, J.-I.; Hayashi, F. ripple current reduction of a fuel cell for a single-phase isolated converter using a DC active filter with a center tap. *IEEE Trans. Power Electron.* **2009**, *25*, 550–556. [CrossRef]
11. Takaoka, N.; Takahashi, H.; Itoh, J.-I. Isolated single-phase matrix converter using center-tapped transformer for power decoupling capability. *IEEE Trans. Ind. Appl.* **2018**, *54*, 1523–1531. [CrossRef]
12. Wang, M.; Huang, Q.; Yu, W.; Huang, A.Q. An isolated bi-directional soft-switched DC-AC converter using wide-band-gap devices with novel carrier-based unipolar modulation technique under synchronous rectification. In Proceedings of the IEEE Applied Power Electronics Conference and Exposition (APEC), Charlotte, NC, USA, 15–19 March 2015; pp. 2317–2324.
13. Varajao, D.; Rui, E.A.; Miranda, L.M.; Lopes, J.A.P.; Weise, N. Control of an isolated sing-phase bidirectional AC-DC matric converter for V2G applications. *Electr. Power Syst. Res.* **2017**, *149*, 19–29. [CrossRef]
14. Norrga, S. Experimental study of a soft-switched isolated bidirectional AC-DC converter without auxiliary circuit. *IEEE Trans. Power Electron.* **2006**, *21*, 6. [CrossRef]
15. Nakata, Y.; Orikawa, K.; Itoh, J.-I. Several-hundred-kHz single-phase to commercial frequency three-phase matrix converter using Delta-sigma modulation with space vector. In Proceedings of the IEEE Energy Conversion Congress and Exposition (ECCE), Pittsburg, PA, USA, 14–18 September 2014; pp. 571–578.
16. Takaoka, N.; Takahashi, H.; Itoh, J.-I.; Chiang, G.T.; Sugiyama, T.; Sugai, M. Power decoupling method comparison of isolated single-phase matrix converters using center-tapped transformer with PDM. In Proceedings of the IEEE Energy Conversion Congress and Exposition (ECCE), Montreal, QC, Canada, 20–24 September 2015.
17. Chiang, G.T.; Takahide, S.; Masaru, S. Optimal design of a matrix converter with a LC active buffer to onboard vehicle battery charger in single phase grid structure. In Proceedings of the 18th European Conference on Power Electronics and Applications, Karlsruhe, Germany, 5–9 September 2016.
18. He, J.; Li, Y.W. Hybrid voltage and current control approach for DG grid interfacing converters with LCL filters. *IEEE Trans. Ind. Electron.* **2013**, *60*, 1797–1809. [CrossRef]
19. She, H.; Lin, H.; He, B.; Wang, X.; Yue, L.; An, X. Implementation of voltage-based commutation in space-vector modulated matrix converter. *IEEE Trans. Ind. Electron.* **2012**, *59*, 154–166.
20. Afsharian, J.; Xu, D.; Wu, B.; Gong, B.; Yang, Z. A new PWM and commutation scheme for one phase loss operation of three-phase isolated buck matrix-type rectifier. *IEEE Trans. Power Electron.* **2018**, *33*, 9854–9865. [CrossRef]

© 2020 by the authors. Licensee MDPI, Basel, Switzerland. This article is an open access article distributed under the terms and conditions of the Creative Commons Attribution (CC BY) license (http://creativecommons.org/licenses/by/4.0/).

Article

Modeling of Magnetic Elements Including Losses—Application to Variable Inductor [†]

Sarah Saeed *, Ramy Georgious and Jorge Garcia

LEMUR Research Group, Department of Electrical, Electronics, Computers and Systems Engineering, University of Oviedo, 33204 Gijon, Spain; georgiousramy@uniovi.es (R.G.); garciajorge@uniovi.es (J.G.)
* Correspondence: saeedsarah@uniovi.es
† This paper is an extended version of our paper published in 2018 IEEE Applied Power Electronics Conference and Exposition (APEC), San Antonio, TX, USA, 4–8 March 2018, pp. 1750–1755.

Received: 4 March 2020; Accepted: 6 April 2020; Published: 11 April 2020

Abstract: This paper proposes and develops a circuit-based model aiming to simulate variable magnetic power elements in power electronic converters. The derived model represents the magnetic element by a reluctance-based equivalent circuit. The model takes into consideration device core losses, with the main emphasis given to hysteresis losses, which are modeled using the Jiles-Atherton model. The core loss model is further validated on different ferromagnetic materials to prove its range of applicability. The winding losses of the magnetic device are also taken into consideration, which are obtained using Dowell empirical formulas. In addition, the frequency dependence of the device losses is also considered. The proposed modeling procedure has been applied to study and characterize a double E-core variable power inductor structure in a 1 kW SiC full bridge DC-DC converter. The procedure has been verified by comparing the simulation results to the experimental measurements, confirming the validity and accuracy of the full circuit-based model.

Keywords: magnetics modeling; variable inductor; hysteresis; eddy currents; saturable core

1. Introduction

Understanding the behavior of a magnetic device in a Power Electronic Converter (PEC) is essential to optimize the design and to foster the performance of the whole system. Variable magnetic elements allow for additional degrees of freedom in the design and control of PECs. This is particularly useful in resonant converters where the usual frequency control has some drawbacks due to Electro-Magnetic Interference (EMI) issues, synchronization, variable sampling time, etc., especially for a large range of variation. If variable magnetics are used, the same control margins can be obtained at a constant switching frequency, therefore allowing for an optimization of the EMI filters and sampling procedures. In other applications, such as the Dual-Active-Bridge (DAB) converter, in addition to adding a new degree of freedom to the control, the inclusion of variable magnetics can increase operation parameters, such as the soft switching margins [1–3].

The recent growing applications of variable magnetic elements have implied the need for developing accurate models to define the magnetic device behavior. The magnetic core as well as the device windings must be characterized to achieve an accurate device model.

Some models define the magnetic core material in terms of a relationship between magnetic flux density and field intensity referred to as a hysteresis curve. In [4], an initial survey has been conducted classifying the existing magnetic material models, according to different frequencies, bias conditions, and temperatures of interest. It aims to provide comparable information for models and their availability in some circuit simulators. More recently, a literature review on the fundamentals, modeling, and design of magnetic regulators has been comprehensively presented in [5]. After a careful review of these and other references, the modeling methods are confined to analytical and numerical methods. Specific to variable magnetic devices, modeling strategies are confined to three directions: Finite Elements Analysis (FEA), gyrator-capacitor model [6], and reluctance equivalent circuit [7]. The FEA model is based on the numerical method, while the gyrator-capacitor model and the reluctance model are based on the analytical method.

As the complexity of magnetic devices increases, the analytical method becomes too complicated to predict the behavior of the device in a simple and practical manner. Therefore, incorporating those concepts in a computer-based simulation provides a good compromise between convenience, accuracy, and numerical efficiency. On the other hand, developing such simulations enables real-time applications on the modeling at the converter controller level, which can provide an on-line calculation of model parameters for real-time control [8]. Consequently, many efforts have been directed towards computer-based simulations, especially time-domain models [9,10]. Although the analytical methods of calculation are generally known and can be implemented in simulation models in a straight-forward manner, the selection of the suitable methods is critical from the circuit simulation perspective. The computation methods are expected to achieve good convergence with an acceptable compromise of accuracy to the time of processing the simulation results.

Henceforth, the aim of this work is to develop a circuit-based time-domain model of the variable magnetic element. The proposed model includes the device losses, mainly core and winding losses. Also, this circuital model is able to work in different platforms, for e.g., LTSpice (Linear Technology Corporation, Milpitas, CA, USA) [11], MATLAB-Simulink (MathWorks, Natick, MA, USA) [12], and PSIM (Powersim Inc., Rockville, MD, USA) [13], with equally valid accuracy. Thereby, the whole electromagnetic system design and simulation can be carried out using only one simulator environment. This provides an acceptable accuracy in compromise with the complication and time required for FEA models. Consequently, it allows the investigation of the overall PEC performance incorporating the variable magnetic device. Section 2 presents an overview of the variable magnetic device structure that will be used, together with the models of interest in the literature. In Section 3, the magnetic core losses are studied and the method used to model those losses is presented. This section also provides an idea on the implementation of the model equations, the validation of the model against experimental measurement, and the approach to estimate the model parameters as a function of the operation frequency. Later, in Section 4, the model of the winding losses is presented, and validated against experimental results. After that, Section 5 explains the use of the loss models to implement the full device model. In Section 6, the proposed simulation model is validated in comparison to the previous models that does not include losses, and experimental results are provided. Finally, Section 7 summarizes the conclusions of the work.

2. Modeling of Variable Magnetic Elements

From the study of the state of the art, the double E-core structure, depicted in Figure 1, is selected to be the most appropriate for the implementation of the variable inductor in this study and the most comprehended in literature [14]. The basic principle of operation of a double E-core variable inductor is described in this section.

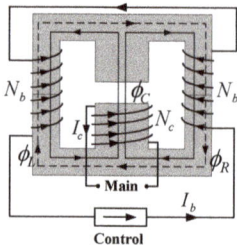

Figure 1. Variable inductor based on a double E-core structure.

Due to the current (I_c) flowing through the main winding (N_c) of the inductor, as clarified in Figure 1, an AC flux (ϕ_C) circulates through the center arm of the typical E-core structure of the magnetic core and splits to the outer arms. Applying a relatively small DC current (I_b) to the bias control windings (N_b), a DC flux (ϕ_R or ϕ_L) is produced, which circulates mainly through the outer (ungapped) closed path of the core [14]. This DC flux can bias the operation of the magnetic material towards the nonlinear region on the $B(H)$ curve, thus causing the inductance seen from the main winding terminals to vary as a function of the DC bias current, Equation (1).

$$L = f(I_b), \tag{1}$$

In order to model the variable magnetic device, the operation regions of the magnetic material on the $B(H)$ curve must be considered. These are divided into three areas: the linear region, the saturation knee, and the non-linear region [9].

In addition, to attain accurate results in the modeling procedure, the different involved device losses are included. There are two main types of losses associated to the magnetic device; core losses and winding losses. For MnZn-ferrite materials, the core losses are composed of three fractions. A fraction of the losses is related to the crystal composition, which is called hysteresis losses. Another fraction is related with the structural form of the core, which is called eddy current losses. As the frequency increases, the eddy currents dominate, and for a frequency range of approximately 1 MHz, those losses can contribute with more than 50% of the total power losses. There is also a third fraction known as resonance losses [15]. This fraction is related to the ferrimagnetic resonance of the material and is pertinent to both the crystal and microstructure.

On the other hand, the winding losses depend on the frequency of operation of the magnetic device. At DC operation, the winding losses are due to the resistance of the copper wire. However, as the frequency of operation increases, two effects take place, which are the skin and the proximity effects. Those effects cause the wire resistance to vary, and this modified resistance is referred to as the AC (or high-frequency) winding resistance.

Although ready-made modules are available in many circuit simulation environments [12,13], it is not considered comprehensive to study the magnetic device. The SPICE-based reluctance equivalent circuit provided by authors in [9] is, thus, found to be of more interest. Figure 2 shows the reluctance equivalent circuit for the double E-core variable inductor.

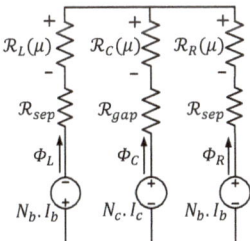

Figure 2. Reluctance equivalent circuit of the double E-core variable inductor shown in Figure 1.

The voltage sources in the circuit represent the magnetomotive forces due to the bias control windings ($N_b \cdot I_b$), as well as the main winding ($N_c \cdot I_c$). $\mathcal{R}_{gap}$ is a constant reluctance that represents the air gap, while $\mathcal{R}_{sep}$ is a constant reluctance that represents the air separation between the two E-cores due to non-idealities in the manufacturing process. The components $\mathcal{R}_L$, $\mathcal{R}_C$, and $\mathcal{R}_R$ are the reluctances of the magnetic paths of the left, center, and right arms, respectively. These reluctance values are represented as a function of the permeability of the magnetic material (μ), and it must be noticed that, unlike the usual case, in variable magnetic elements, these permeability values can vary depending on the operation point of the magnetic core material on the characteristic $B(H)$ curve. Thus, for calculating these reluctances, the referred model uses Brauer's equation [16], which defines the $B(H)$ characteristic curve of the magnetic material neglecting the hysteresis effect, as stated by Equation (2).

$$H(B) = (k_1 e^{k_2 B^2} + k_3) B, \qquad (2)$$

where k_1, k_2, and k_3 are constants that depend on the considered magnetic material. For the present work, the afore-mentioned SPICE-based model has been replicated, specifically, in the MATLAB-Simulink platform. This has been carried out in order to take advantage of the feature of this environment to integrate MATLAB script with existing Simulink library tools, aiming to combine complex, accurate simulations with digital processing of information. This allows including the device design calculations into the overall model of the system in one integrated environment.

Two key limitations are found in this model, which restrict its applicability range and accuracy. Firstly, the hysteresis effect is not taken into consideration. Secondly, the model input quantities are dependent on the output ones, which introduces difficulty in the implementation of the computations. In particular, the latter issue implies the necessity for implementing system calculation delays, with an adequate initialization of parameters, especially when including the device in a switching converter simulation. This, furthermore, complicates the simulation in different test platforms. In Simulink, for instance, the solver applies a numerical method to solve the set of ordinary differential equations that represent the model; therefore, the model causality must be decided [17].

In this paper, the model of the variable inductor has been extended to include device losses. The main loss components that have been taken into consideration are the hysteresis losses of the magnetic core and the winding losses. The following sections will justify the studied loss components and discuss the implementation and validation of the full model [18].

It is worth noting that, throughout the discussion hereafter, the vectorial nature of the magnetic quantities are disregarded to reduce the analysis to a simple unidimensional statement of equations by assuming: (1) specific symmetrical magnetic core geometries, of which geometrical references and paths are well-defined, and (2) the homogeneous nature of typical magnetic core materials together with the uniform distribution of the magnetic core properties. Such assumptions are often used in the analysis and design of magnetic devices for power electronics converters.

3. Model of Core Losses

The core losses in a magnetic core are due to two phenomena: the hysteresis loss, and the eddy currents loss. The hysteresis loss is due to energy required to rotate the magnetic domains when aligning with the applied magnetic field. On the other hand, the eddy current losses are due to currents induced in the magnetic core, which opposes the changing flux in the core. Previous studies in literature [19] have shown that for a ferrite magnetic material operating in a range of frequency up to 100 kHz, the eddy current losses are a very small part of the total core losses. Therefore, for the range of frequencies under study herein, the hysteresis losses will always be dominant. For this reason, the eddy current losses in the magnetic core will be neglected in this study for the sake of simplicity.

There are several methods to calculate the hysteresis losses in a ferro-magnetic material, which were grouped by the authors in [20] to be three main approaches: hysteresis models, empirical equations, and loss separation. This paper undertakes the first approach, specifically the Jiles-Atherton (JA) hysteresis model [21]. The main strengths of the JA model compared to its counterpart approaches are: being the most suitable for development from a circuital simulation perspective, besides having good convergence and acceptable accuracy among a variety of materials and operation conditions [22].

This section is dedicated to explaining in detail the model of core losses. First, the JA equations are defined and implemented using block diagram modeling in Simulink. Second, a test setup is built for measuring the magnetic flux density and field intensity to validate the model in comparison to experimental measurements. Finally, to include the effect of the switching frequency on the core loss model, expressions are obtained for the JA parameters as a function of the switching frequency based on an empirical approach.

3.1. Model Implementation

The JA model separates the magnetization, M, into reversible, M_{rev}, and irreversible magnetizations, M_{irr}, which correspond to the reversible or irreversible phenomena, which take place within the magnetic material during the magnetization [22]. By computing the latter components, the total magnetization, M, is defined as the summation of both components as explained by Equation (3).

$$M = M_{irr} + M_{rev}. \tag{3}$$

The reversible component of magnetization is defined as a fraction, c, of the difference of the anhysteretic magnetization and the irreversible one, as explained by Equation (4).

$$\frac{dM_{rev}}{dH} = c \left(\frac{dM_{an}}{dH} - \frac{dM_{irr}}{dH} \right), \tag{4}$$

where c is referred to as the reversibility coefficient. The irreversible component of magnetization is defined by the differential Equation (5).

$$\frac{dM_{irr}}{dH} = \frac{M_{an} - M_{irr}}{\frac{\delta k}{\mu_0} - \alpha (M_{an} - M_{irr})}, \tag{5}$$

where k is the loss coefficient, and α is the interdomain coupling. δ indicates the direction of the magnetizing field H, such that $\delta = 1$ for increasing field, and $\delta = -1$ for decreasing field, as defined by Equation (6).

$$\delta = \begin{cases} 1, & \frac{dH}{dt} > 0. \\ -1, & \frac{dH}{dt} < 0. \end{cases} \tag{6}$$

M_{an} is the anhysteretic magnetization. The anhysteretic magnetization describes the magnetization of an ideal ferromagnet that does not have a loss effect, and thus, its magnetization curve does not present hysteresis. Langevin's function [23] is used within the JA model for defining the anhysteretic magnetization, in which case, an effective field, H_e, replaces the magnetic field, H, as explained by Equation (7) [24].

$$M_{an} = M_s \left(\coth \frac{H_e}{a} - \frac{a}{H_e} \right), \tag{7}$$

where M_s is the saturation magnetization, a is the shape parameter for anhysteretic magnetization, and H_e is defined by Equation (8).

$$H_e = H + \alpha M. \tag{8}$$

Consequently, the magnetic flux density can be calculated using Equation (9).

$$B = \mu_0 H_e. \tag{9}$$

Table 1 summarizes the definition of the model parameters, M_s, a, c, α and k. The parameters are initially estimated by an iterative procedure to fit the model to the magnetic material $B(H)$ curve data provided by the manufacturer.

Table 1. Jiles-Atherton model parameters.

M_s	Saturation magnetization	A/m
a	Shape parameter for anhysteretic magnetization	A/m
k	Pinning (or loss) coefficient	A/m
c	Reversibility coefficient	—
α	Interdomain coupling	—

Figure 3 shows a block diagram of the detailed implementation of the JA equations [22]. Following this block diagram, the model equations have been implemented in Simulink. Therefore, for a given core size and magnetic material, the instantaneous magnetic flux density (B) can be estimated for a certain instantaneous magnetic field intensity (H) applied to the magnetic core.

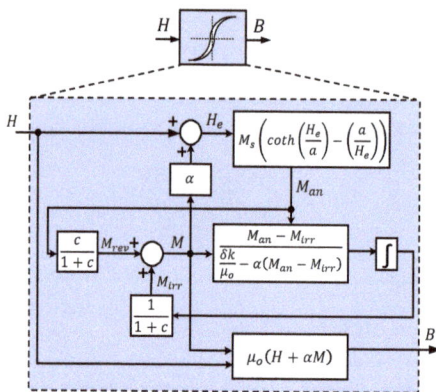

Figure 3. Schematic of the implementation of the Jiles-Atherton (JA) model.

3.2. Test Setup for Measuring Core Losses

In order to validate the implemented JA model and study the effect of frequency on the model parameters, a test setup is developed. The purpose of the experiments is to measure the hysteresis losses of the magnetic core and compare it to the one obtained by the implemented JA model. In literature, mainly two approaches for measuring core losses can be found [25]: electrical methods, and calorimeter-based methods. One of the former approaches has been selected, which is the $B(H)$ curve electrical measurement technique. Specific to the selected technique, the two-winding measurement method [26] is used since it is reported to be accurate for the frequency range under test in this study (<100 kHz) [27].

Figure 4 illustrates a schematic to clarify the test setup used for measuring the core losses, while Figure 5 shows the experimental platform developed.

The power stage used in the tests is a simple half-bridge converter suitable for low power levels. A square-waveform excitation voltage is sought to verify the loss study under non-sinusoidal conditions. The core used for the identification tests is a toroidal core with the design parameters listed in Table 2. This toroidal geometry is used in order to ensure that the model results are not only valid for the selected double E-core structure but also for other geometrical schemes.

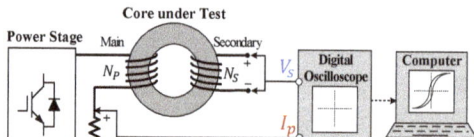

Figure 4. Schematic diagram of the test setup used to measure the $B(H)$ curve.

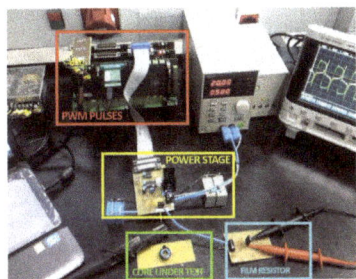

Figure 5. Experimental setup used to measure the $B(H)$ curve.

Table 2. Specifications of the test setup developed to measure the $B(H)$ curves.

Magnetic Core	
Core material	N87
Core type and size	Toroidal core: R16.0 × 9.60 × 6.30
Main winding no. of turns (N_p)	5 turns
Sensing winding no. of turns (N_s)	5 turns
Power Stage Ratings	
Power level	100 W
Input voltage	30 V
Peak current	3 A
Frequency (f)	50 kHz

In addition to the main excitation winding (N_p), a secondary sensing winding (N_s) is added to sense the induced voltage due to the flux in the main one. The advantage of a separate winding is to exclude the voltage drop due to the resistance of the main winding. Therefore, the magnetic flux density can be computed using Equation (10).

$$B = \frac{1}{N_s A} \int_0^T V_s dt, \tag{10}$$

where A is the cross-section area of the toroid, and V_s is the open-circuit secondary winding voltage. The integration is performed over one switching period, T. The field intensity can, thus, be computed using Equation (11).

$$H = \frac{N_p I_p}{l}, \tag{11}$$

where I_p is the current flowing through the main winding, and l is the effective length of the core. The measured voltage (V_s) and current (I_p) are illustrated in Figure 6a,b, respectively. By applying these measurements to Equations (10) and (11), the B and H quantities can be calculated. In this manner, the measured $B(H)$ curve of the N87 magnetic material is compared to the $B(H)$ curve obtained from the implemented JA model simulation under the same operation conditions. To apply the JA model, the model parameters are estimated by iterative fitting as mentioned previously, the obtained parameters for N87 magnetic material at 50 kHz are stated in Table 3. The resulting comparison is shown in Figure 7. It can be observed that the JA simulation model predicts the $B(H)$ curve of the N87 magnetic material with an error of less than 5%. It can, thus, be concluded that the model is valid and provides acceptable accuracy. Furthermore, the model has been tested under different magnetic materials. The previous measurements were repeated at the exact same operation conditions while using a similar toroidal core from 3C90 magnetic material, the JA parameters of which are stated in Table 3. The B and H quantities were again measured, and the $B(H)$ curves were compared together with the modeled ones, as illustrated in Figure 7. Similar to the previous results, the model shows a clear coincidence with the experimental measurements.

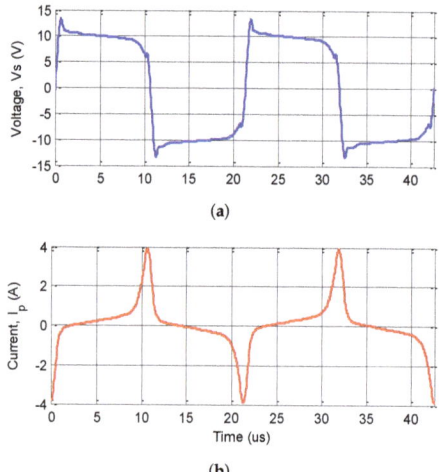

Figure 6. Experimental measurements. (**a**) Voltage applied to the toroid, and (**b**) current flowing through the main winding.

Table 3. Estimated Jiles-Atherton parameters for N87 and 3C90 magnetic materials at 50 kHz.

JA Parameter	N87	3C90
M_s	$4.0481e^5$	$3.7547e^5$
a	17.7019	19.5349
k	12.5883	12.8057
c	0.3210	0.3210
α	$2.0000e^{-5}$	$2.0000e^{-5}$

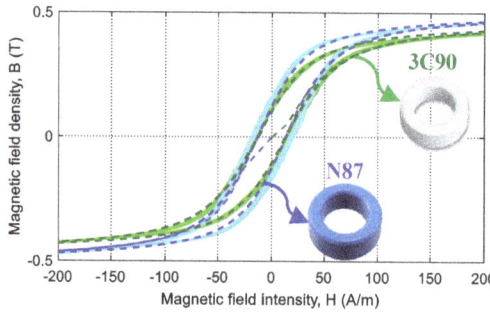

Figure 7. Model (dotted line) versus experimental (solid line) $B(H)$ curves for two different magnetic materials: N87 (blue) and 3C90 (green).

The hysteresis losses are estimated in terms of the area enclosed by the $B(H)$ loop and, therefore, can be expressed as shown by Equation (12).

$$P_{core} = V_e f \oint B(H) dH, \tag{12}$$

where V_e is the core volume, and f is the switching frequency. The integral is performed over a complete cycle of the magnetic field intensity.

3.3. Frequency Dependence of JA Parameters

Since the behavior of the magnetic core material depends on the excitation frequency, different voltage waveforms in the magnetic element imply notable variations in the associated trajectories on the $B(H)$ characteristic of the material. Specifically, it changes the area enclosed in the hysteresis loop, which is directly related to the core losses. This variation of the trajectory in itself implies a variation of the parameters of the JA model parameters. It is, thus, of interest to develop a model that can be used for any frequency without having to readjust the parameters each time different operating conditions are considered.

A practical approach has been taken by conducting a few experiments to characterize the variation of the JA model parameters as a function of the frequency. For the same core size and material, a number of independent tests were carried out, each test for a given frequency of operation. Then, a simple procedure is followed to obtain the parameters, as listed below:

1. The instantaneous waveforms of the voltage (V_s) and the current (I_p), as explained in the previous section, are captured.
2. Fast Fourier Transform (FFT) is applied to these waveforms in order to identify the waveform frequency.
3. For each switching frequency test, the $B(H)$ curve is measured using Equations (10) and (11). Then the measured $B(H)$ curve is fitted to the modeled one by using iterative trial-and-error steps in the JA parameters.
4. For each test, the set of obtained JA model parameters is collected.
5. Using curve fitting techniques (specifically the Curve Fitting Toolbox from Matlab), an expression is obtained for each of the JA model parameters $M_s(f)$, $a(f)$, and $k(f)$ as a function of switching frequency. The parameters c and α show insignificant variation with frequency.

The obtained expressions for the JA parameters are stated by Equations (13)–(15).

$$M_s(f) = 5.189e^{-8}f^{2.334} + 4e^5, \tag{13}$$

$$a(f) = 6.004e^{-15}f^{3.002} + 16.935, \tag{14}$$

$$k(f) = -3.398e^{-7}f^{1.458} + 15. \tag{15}$$

The expressions are intrinsic to a certain magnetic material. In this case, the procedure has been implemented for the N87 material; however, the same procedure can be followed for extracting a set of expressions for the JA parameters of any other ferrite material.

To test the validity of the obtained expressions for the JA parameters as a function of frequency, several experiments were carried out using the N87 toroidal core described in Table 2. The switching frequency of the converter was changed and the $B(H)$ curves corresponding to different frequency values were measured and compared to the modeled ones, as illustrated in Figure 8. The results of the comparison show that the modeled $B(H)$ curves match the obtained experimental measured curves at a wide range of the operation frequency with acceptable accuracy.

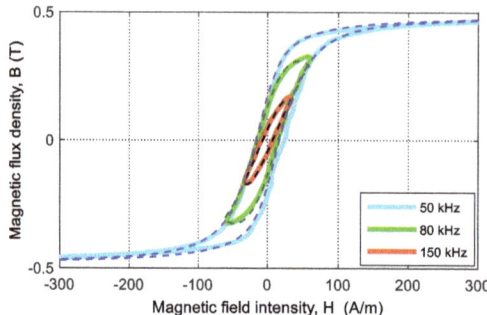

Figure 8. $B(H)$ curves for the prototype under different operation frequencies. Model (dotted line) versus experimental results (solid line).

4. Model of Winding Losses

4.1. Winding Eddy Current Losses

The winding losses are due to the resistance of the copper wire. At DC operation currents or relatively low operation frequencies, this resistance component is constant and calculated using Equation (16).

$$R_{dc} = \rho_{cu} \frac{l_{wire}}{A_{wire}}, \tag{16}$$

where ρ_{cu} is the resistivity of copper material at 20°C, which is equal to 1.68×10^{-8} Ωm, l_{wire} is the total length of the winding wire, and A_{wire} is the cross-section area of the wire.

However, as the switching frequency increases, two effects start to appear, which are the skin effect and proximity effect. These effects induce eddy currents in the winding conductors, altering the resistance of the winding, and significantly contributing to the overall winding losses. It is necessary in this case to calculate the AC resistance of the winding. Dowell provided a method that computes the equivalent winding resistance using a one-dimensional analytical approach [28]. Initially, this method was intended to describe high-frequency loss in foil windings; however, it has been extended to multilayer windings with round conductors by introducing the porosity factor [28,29]. Dowell's method uses a sinusoidal approach, and the calculations are limited to non-gapped cores [30]. On the other hand, the method presents a great advantage of simplicity and a fast computation of the AC winding resistance; thus, it can be easily integrated into the magnetic device model without extra complexity. Dowell estimates the AC resistance (R_{ac}) by scaling the DC winding resistance by a factor, as shown in Equation (17).

$$R_{ac} = R_{dc} \left(M' + \frac{(m^2 - 1) D'}{3} \right), \tag{17}$$

where M' and D' are coefficients defined based on the geometrical dimensions of the winding, material characteristics, and frequency of operation, and m is the number of layers. The accuracy of the full winding model will be provided in the following sections to assess the validity of using Dowell's method for the application herein.

4.2. Winding Stray Capacitance

As the operation frequency increases, the parasitic capacitance of the inductor winding becomes more significant, causing the impedance of the inductor to change and introduce the resonant frequencies. In order to attain a full comprehensive model of the device, the stray capacitance of the windings is added to the model. The calculation of the stray capacitance is based on an analytical approach previously presented in literature [31]. This method is valid for multi-layer inductors with ferromagnetic cores, as well as being simple and reliable for simulation purposes. Briefly, the inductor winding is divided into partitions, and the turn-to-turn and turn-to-core capacitances of the winding are predicted as a function of a few geometrical parameters of the device. Accordingly, the overall stray capacitance of the coil (C_s) converges to the expression stated in Equation (18).

$$C_s \cong 1.366 C_{tt}, \tag{18}$$

where C_{tt} is the turn-to-turn capacitance of the coil and is defined by Equation (19).

$$C_{tt} = \varepsilon_0 l_t \left(\frac{\varepsilon_r \theta^*}{\ln \frac{D_o}{D_c}} + \cot\left(\frac{\theta^*}{2}\right) - \cot\left(\frac{\pi}{12}\right) \right), \tag{19}$$

where l_t is the turn length, θ^* is the angular coordinate, ε_0 and ε_r are the permittivity of air and relative permittivity of the insulation medium, respectively, and D_o and D_c are the diameters of the wire with and without the insulation coating, respectively, as clarified by Figure 9.

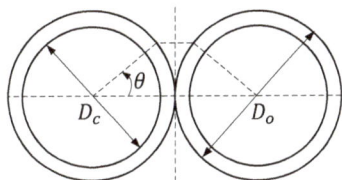

Figure 9. Two adjacent turns of the winding.

4.3. Full Winding Model

The full model of the inductor winding is expressed by the circuit diagram in Figure 10. Therefore, the total impedance of the winding is calculated by Equation (20).

$$Z_T = \frac{(sL + R_{ac} + R_{dc}) \frac{1}{sC_s}}{sL + R_{ac} + R_{dc} + \frac{1}{sC_s}}. \tag{20}$$

An inductor prototype was implemented based on the specifications summarized in Table 4. The inductance value is not of specific importance in the design; however, it is interesting to distribute the winding on several layers to emphasize the proximity effect. Figure 11 illustrates the total winding impedance as a function of frequency to compare the developed winding model against the experimental measurements. As it can be observed, the error between the measured and modeled impedances is less than 1%, which represents a quite high accuracy for the study in context.

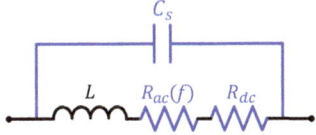

Figure 10. Full winding model.

Table 4. Specifications of the inductor prototype developed to validate the winding model.

Magnetic Core	N87
Core shape and size	ETD49
Main winding no. of turns	80 turns
No. of winding layers, m	4 layers
Diameter of wire, D_{wire}	1 mm

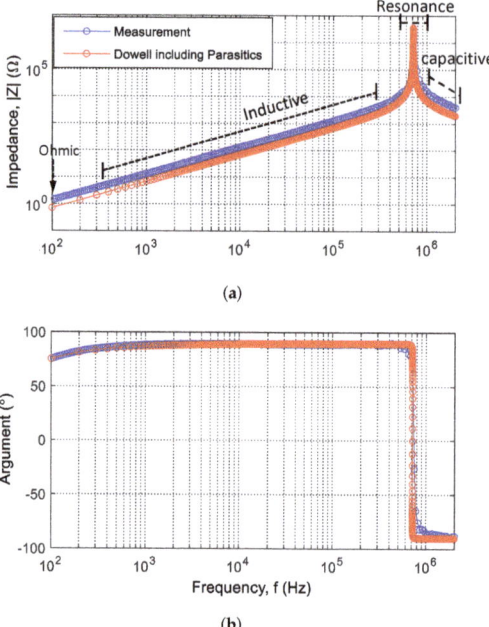

Figure 11. Total winding impedance as a function of frequency. (a) Magnitude, and (b) argument of modeled impedance compared to experimental measurements.

5. Applications of Loss Models to Simulate the Variable Inductor

After the verification of the loss models, the study is applied to model the double E-core variable inductor shown in Figure 1. The full model is implemented based on the reluctance circuit concept previously stated. However, the two issues associated with the previous reluctance model have been tackled. To avoid algebraic loops due to model causality, the magnetic core has been partitioned according to its operation, as mentioned previously, in Section 2, thus defining each partition in terms of electrical inputs and outputs as explained below.

The middle arm has the main winding, and to model this arm, the input will be the excitation voltage, and the output should be the main inductor current. On the other hand, the lateral arms have the control windings, so for these coils, the input will be the control current, and the output should be the induced voltages in the control windings.

Figure 12a illustrates the magnetic system, which is represented by the reluctance equivalent circuit of the device. The reluctance equivalent circuit is composed of three branches. The left and right branches represent the magnetic circuits of the control arms of the device. The voltage source ($N_b \cdot I_b$) models the magnetomotive force created by each control winding. The voltage sources ($\phi_R \cdot \mathfrak{R}_R$) and ($\phi_L \cdot \mathfrak{R}_L$) model the variable reluctance of the magnetic path of the right and left arms, respectively. Using the control current as the input quantity, the values of the variable voltage sources are calculated based on the JA hysteresis model. The middle branch represents the magnetic circuit of the main arm, the variable reluctance of the magnetic path is similarly represented by the voltage source ($\phi_C \cdot \mathcal{R}_C$), while in this case, the magnetomotive force is the output quantity and is represented by a current source. The current in the main winding (I_c) can, thus, be calculated by measuring the voltage across this current source and dividing by the number of turns of the winding (N_c).

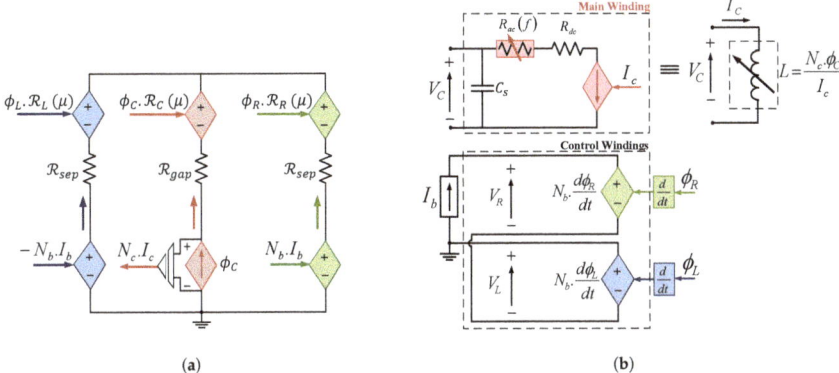

Figure 12. Schematic of the variable inductor model based on the reluctance circuit. (a) Magnetic circuit, and (b) electric circuit.

On the other hand, Figure 12b illustrates the electrical system, which includes the electrical model of the winding, as well as the input voltage and the output current source. The electrical part of the three windings is represented by Equation (21).

$$V_w = N_w \frac{d\phi_w}{dt}, \quad w \in L, R, C \qquad (21)$$

where V_w is the voltage at winding w, ϕ_w is the magnetic flux created by the winding w, and N_w is the number of turns of the winding w. The winding w refers to the left (L), right (R), or center (C) windings. In order to account for the main winding DC losses, a constant resistor, R_{dc}, is added to the electrical circuit of the main winding in series with the current-controlled current source, which represents the main winding current. Also, to represent the AC winding losses, a variable frequency-dependent resistor, $R_{ac}(f)$, is added in series to R_{dc}.

6. Model Validation Using Simulations and Experimental Results

In order to compare the initial lossless equivalent circuit with the proposed model, which includes core and winding losses, detailed simulations have been carried out. Furthermore, those two simulation models have been compared against experimental measurements obtained from the variable inductor prototype shown in Figure 13. The device has been developed based on the double E-core structure with the design specifications indicated in Table 5, and the models have been adjusted correspondingly.

Figure 13. Variable inductor prototype based on double the E-core structure.

Table 5. Specifications of the test setup developed to validate the full VI model.

Magnetic Core	
Core material	N87
Core shape and size	ETD49/25/16
Main winding no. of turns	23 turns
Control winding no. of turns	55 turns
Power Stage Ratings	
Power level	1 kW
Input voltage	200 V
Peak current	5 A
Frequency (f)	50 kHz

Figure 14a shows a circuit diagram of the developed test platform. It consists of a SiC full-bridge DC-AC converter to apply a square waveform excitation voltage on the inductor main winding. As mentioned previously, the square waveform voltage allows testing the device model under non-sinusoidal conditions, thus assure the validation of the loss study under a general condition of excitation voltage. The converter is controlled using a TMS320F28335 Texas Instruments peripheral board. Additionally, a variable DC voltage source is connected in series with a resistor to provide a DC control current of maximum 1 A to the control winding of the variable inductor. The constructed test platform is illustrated in Figure 14b.

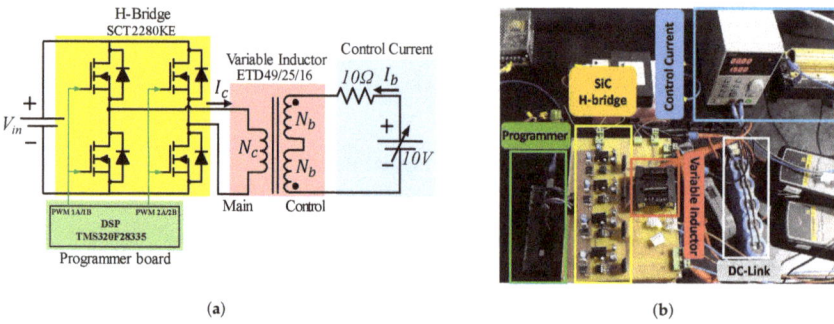

Figure 14. Experimental setup used to test the variable inductor under small and large-signal analyses. (**a**) Circuit diagram, and (**b**) test platform.

6.1. Small-Signal Analysis

To validate the proposed model under small-signal analysis, the control current was increased from 0 to 1 A in steps of 0.05 A, while keeping zero excitation voltage on the main winding. The equivalent inductance seen from the main winding was measured using the impedance analyzer. Figure 15a illustrates the measurement of the equivalent inductance as a function of the bias control current. Also, the figure illustrates the simulated inductance obtained from the developed models, the initial lossless equivalent circuit, as well as the proposed model, which includes losses. Figure 15b shows the error of each model compared to the experimental results, as calculated by Equation (22). It can be observed that the proposed model that includes losses predicts the inductance within an acceptable error (<6%). On the other hand, the inductance predicted by the initial lossless model shows a clear deviation from the experimental one as the control current increases. It reaches an error of 30% at maximum control current.

$$Error(\%) = \frac{(L_{Experiment} - L_{Model})}{L_{Experiment}} \times 100. \tag{22}$$

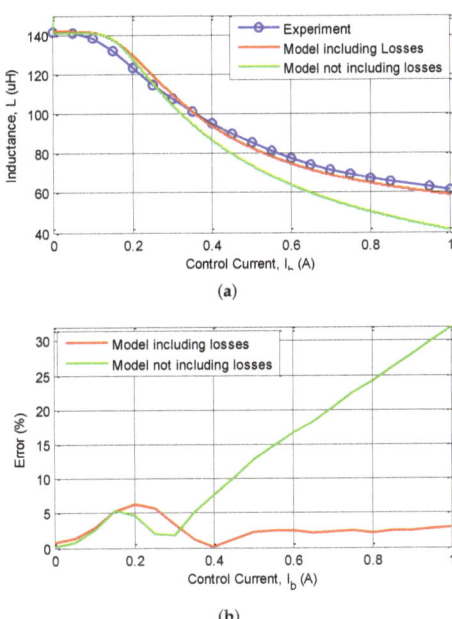

Figure 15. Small-signal characterization of the variable inductor prototype comparing simulation models with experimental results. (**a**) Inductance value, and (**b**) percentage error as a function of control current.

6.2. Large-Signal Analysis

The prototype has also been characterized under large-signal analysis. Similar to the small-signal analysis, a DC control current is applied to the control windings and varied from 0 to 1 A. However, in this case, a square waveform voltage of 30 V is applied to the main winding of the inductor. The inductance is calculated by two different methods using the experimental measurements of the voltage and current through the main winding. The first method calculates the inductance using the RMS values of the waveforms over each cycle at a steady state. On the other hand, the second method uses the instantaneous values of the waveforms. The two methods were then compared to verify the accuracy of the measured inductance value, as explained hereafter.

6.2.1. Impedance Calculation

A simplification applied by considering the RMS value of the first harmonic component of the voltage and current measurements, so the resulting inductance is calculated by Equation (23).

$$L = \frac{X_L}{\omega} = \frac{V_C}{I_C} \cdot \frac{\pi/4}{2\pi f}, \qquad (23)$$

where V_C is the voltage applied to the main winding of the variable inductor, and I_C is the current flowing through it.

6.2.2. System Identification Tools

The System Identification Toolbox from Matlab is used, which requires a set of data that represent the input and output variables of a system; in this case, V_C is the input to the system, and I_C is the output. Using these variables, the tool defines the transfer function of the system based on the user selection of the number of poles and zeros of the system. In the case of an inductor, the system is defined as a 1st order system (1 pole, no zeros), and the transfer function is stated by Equation (24).

$$\frac{I_C(s)}{V_C(s)} = \frac{1}{sL + R_{dc}}. \qquad (24)$$

Figure 16 compares both methods of inductance calculation to verify the accuracy of the measured inductance and then uses it to assess the proposed model. It shows that the inductance calculated based on RMS values is in very close agreement with the inductance obtained by the Matlab identification tool. This conclusion justifies the use of RMS measurements for calculating the inductance in the large-signal analysis in order to simplify the computations.

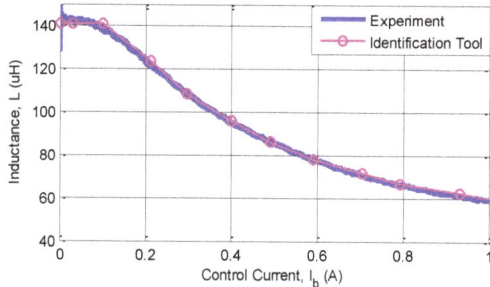

Figure 16. Inductance calculation from experimental measurement.

Similar to the previous small-signal analysis, Figure 17a illustrates the equivalent inductance as a function of the bias control current for the large-signal analysis case. Figure 17b clarifies the trend of deviation or convergence of the models as a function of the control current, compared to the measured results. The results are quite consistent with the previous small-signal analysis; the inductance calculated by the proposed model still approaches the experimental measurements, while the one calculated by the lossless model shows a clear dispersion.

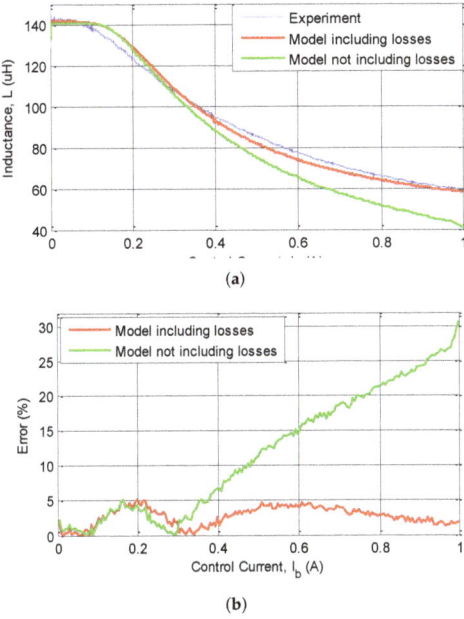

Figure 17. Large-signal characterization of the variable inductor prototype comparing simulation models with experimental results. (**a**) Inductance value, and (**b**) percentage error as a function of control current.

7. Conclusions and Future Developments

In this paper, a full accurate circuit-based time-domain model for a variable magnetic device has been developed, demonstrated, and experimentally validated. The model can predict the inductance variation as a function of the control current in a variable magnetic element. The main contribution of this work is the development of a model that can be used in several simulation platforms, and, moreover, the inclusion of core losses as well as winding eddy current losses in the magnetic device. Additionally, it solves the causality issues, which are present in previous approaches.

A double E-core variable inductor prototype has been characterized under small-signal as well as large-signal analyses in order to assess the accuracy of the model. The proposed approach is considered an autonomous tool that can analyze any given set of data (simulated or experimental) for a magnetic core, detect the operation frequency, and correspondingly, adjust the magnetic core model with the core and winding loss parameters, and finally predict the inductance as a function of the control current, along with other electric and magnetic quantities that characterize the magnetic core operation. Consequently, only one simulator environment is used for the design and simulation of the electromagnetic system.

The test results validate the model under operation frequencies of hundreds of kHz (<200 kHz). Accordingly, the future developments of this work include the extension of the model's validity to frequency ranges of 1–1.5 MHz. Also, the employment of the developed electromagnetic model to study the behavior of the variable inductor in a power electronic converter, specifically a DAB converter. The proposed model will allow for several studies using time-domain simulations, such as the control of the converter power transfer using the variable inductor, the possibility of the linearization of the system transfer function, and finally, the boost of the efficiency over critical operation ranges, for example, light load and heavy load operation ranges.

Author Contributions: S.S. and J.G. conceived the research and designed and performed the experiments; R.G. contributed to the reviewing and editing; all the authors analyzed the data and contributed in the discussion and conclusions. All authors have read and agreed to the published version of the manuscript.

Funding: This work has been partially supported by the Spanish Government, Innovation Development, and Research Office (MEC), under research grant ENE2016-77919, Project "Conciliator", and by the European Union through ERFD Structural Funds (FEDER). Also, this work has been partially supported by the government of Principality of Asturias, Foundation for the Promotion in Asturias of Applied Scientific Research and Technology (FICYT), under Grant FC-GRUPIN-IDI/2018/000241 and under Severo Ochoa research grants, PA-13-PF-BP13-138 and PF-BP16-133.

Conflicts of Interest: The authors declare no conflict of interest. The founding sponsors had no role in the design of the study; in the collection, analyses, or interpretation of data; in the writing of the manuscript, and in the decision to publish the results.

Abbreviations

The following abbreviations are used in this manuscript:

PEC	Power Electronic Converter
EMI	Electro-Magnetic Interference
DAB	Dual-Active-Bridge
FEA	Finite Elements Analysis
JA	Jiles-Atherton
FFT	Fast Fourier Transform

References

1. Fan, H.; Li, H. High-Frequency Transformer Isolated Bidirectional DC–DC Converter Modules With High Efficiency Over Wide Load Range for 20 kVA Solid-State Transformer. *IEEE Trans. Power Electron.* **2011**, *26*, 3599–3608. [CrossRef]
2. Burgio, A.; Menniti, D.; Motta, M.; Pinnarelli, A.; Sorrentino, N.; Vizza, P. A laboratory model of a dual active bridge DC-DC converter for a smart user network. In Proceedings of the 2015 IEEE 15th International Conference on Environment and Electrical Engineering (EEEIC), Rome, Italy, 10–13 June 2015.
3. Saeed, S.; Garcia, J. Extended Operational Range of Dual-Active-Bridge Converters by using Variable Magnetic Devices. In Proceedings of the 2019 IEEE Applied Power Electronics Conference and Exposition (APEC), Anaheim, CA, USA, 17–21 March 2019; pp. 1629–1634.
4. Takach, M.; Lauritzen, P. Survey of magnetic core models. In Proceedings of the 1995 IEEE Applied Power Electronics Conference and Exposition, Dallas, TX, USA, 5–9 March 1995.
5. Chen, Q.; Xu, L.; Ruan, X.; Wong, S.C.; Tse, C.K. Research and Development on New Control Techniques for Electronic Ballasts Based on Magnetic Regulators. Ph.D. Thesis, Dept. Elect. Comput. Eng., Univ. Coimbra, Coimbra, Portugal, 2012.
6. Chen, Q.; Xu, L.; Ruan, X.; Wong, S.C.; Tse, C.K. Gyrator-Capacitor Simulation Model of Nonlinear Magnetic Core. In Proceedings of the 2009 Twenty-Fourth Annual IEEE Applied Power Electronics Conference and Exposition, Washington, DC, USA, 15–19 February 2009.
7. Ludwig, G.; El-Hamamsy, S.A. Coupled inductance and reluctance models of magnetic components. *IEEE Trans. Power Electron.* **1991**, *6*, 240–250. [CrossRef]
8. Almaguer, J.; Cárdenas, V.; Espinoza, J.; Aganza-Torres, A.; González, M. Performance and Control Strategy of Real-Time Simulation of a Three-Phase Solid-State Transformer. *Appl. Sci.* **2019**, *9*, 789. [CrossRef]
9. Alonso, J.M.; Martinez, G.; Perdigao, M.; Cosetin, M.; do Prado, R.N. Modeling magnetic devices using SPICE: Application to variable inductors. In Proceedings of the 2016 IEEE Applied Power Electronics Conference and Exposition (APEC), Long Beach, CA, USA, 20–24 March 2016.
10. Mandache, L.; Topan, D.; Sirbu, I.G. Accurate Time-Domain Simulation of Nonlinear Inductors Including Hysteresis and Eddy-Current Effects. In Proceedings of the World Congress on Engineering, London, UK, 6–8 July 2011; Volume 2.
11. LTspice Design Center. Available online: https://www.analog.com/en/design-center/design-tools-and-calculators/ltspice-simulator.html (accessed on 10 April 2020).

12. MathWorks Simulink Documentation. Available online: https://es.mathworks.com/help/simulink/index.html?s_cid=doc_ftr (accessed on 10 April 2020).
13. PSIM User's Guide. Available online: https://www.myway.co.jp/products/psim/dlfiles/pdf/PSIM_User_Manual_V9.0.2.pdf (accessed on 10 April 2020).
14. Medini, D.; Ben-Yaakov, S. A current-controlled variable-inductor for high frequency resonant power circuits. In Proceedings of the 1994 IEEE Applied Power Electronics Conference and Exposition, Orlando, FL, USA, 13–17 February 1994.
15. Sun, J. Recent Development in Ferrite Material for High Power Application [Passive Components]. *IEEE Power Electron. Mag.* **2018**, *5*, 21–25. [CrossRef]
16. Brauer, J. Simple equations for the magnetization and reluctivity curves of steel. *IEEE Trans. Magn.* **1975**, *11*, 81–81. [CrossRef]
17. Johansson, P.; Andersson, B. Comparison of Simulation Programs for Supercapacitor Modelling. Master of Science Thesis, Chalmers University of Technology, Gothenburg, Sweden, 2008. Available online: http://webfiles.portal.chalmers.se/et/MSc/AnderssonJohanssonMSc.pdf (accessed on 10 April 2020).
18. Saeed, S.; Garcia, J.; Georgious, R. Modeling of variable magnetic elements including hysteresis and Eddy current losses. In Proceedings of the 2018 IEEE Applied Power Electronics Conference and Exposition (APEC), San Antonio, TX, USA, 4–8 March 2018.
19. Cardelli, E.; Fiorucci, L.; Torre, E.D. Estimation of MnZn ferrite core losses in magnetic components at high frequency. *IEEE Trans. Magn.* **2001**, *37*, 2366–2368. [CrossRef]
20. Reinert, J.; Brockmeyer, A.; Doncker, R.D. Calculation of losses in ferro- and ferrimagnetic materials based on the modified Steinmetz equation. *IEEE Trans. Ind. Appl.* **2001**, *37*, 1055–1061. [CrossRef]
21. Jiles, D.; Atherton, D. Ferromagnetic hysteresis. *IEEE Trans. Magn.* **1983**, *19*, 2183–2185. [CrossRef]
22. Wilson, P.R. Modelling and Simulation of Magnetic Components in Electric Circuits. Ph.D. Thesis, School of Electronics and Computer Science, University of Southampton, Southampton, UK, 2001. Available online: https://eprints.soton.ac.uk/368484/ (accessed on 10 April 2020).
23. Ramesh, A.; Jiles, D.; Roderick, J. A model of anisotropic anhysteretic magnetization. *IEEE Trans. Magn.* **1996**, *32*, 4234–4236. [CrossRef]
24. Pop, N.; Caltun, O. Jiles-Atherton Magnetic Hysteresis Parameters Identification. *Acta Phys. Pol. A* **2011**, *120*. Available online: http://przyrbwn.icm.edu.pl/APP/PDF/120/a120z3p22.pdf (accessed on 10 April 2020). [CrossRef]
25. Xiao, C.; Chen, G.; Odendaal, W.G.H. Overview of Power Loss Measurement Techniques in Power Electronics Systems. *IEEE Trans. Ind. Appl.* **2007**, *43*, 657–664. [CrossRef]
26. Thottuvelil, V.; Wilson, T.; Owen, H. High-frequency measurement techniques for magnetic cores. *IEEE Trans. Power Electron.* **1990**, *5*, 41–53. [CrossRef]
27. Mu, M.; Li, Q.; Gilham, D.J.; Lee, F.C.; Ngo, K.D.T. New Core Loss Measurement Method for High-Frequency Magnetic Materials. *IEEE Trans. Power Electron.* **2014**, *29*, 4374–4381. [CrossRef]
28. Dowell, P. Effects of eddy currents in transformer windings. *Proc. Inst. Electr. Eng.* **1966**, *113*, 1387. [CrossRef]
29. Yin, Y.; Li, L. Improved method to calculate the high-frequency eddy currents distribution and loss in windings composed of round conductors. *IET Power Electron.* **2017**, *10*, 1494–1503. [CrossRef]
30. Holguin, F.A.; Asensi, R.; Prieto, R.; Cobos, J.A. Simple analytical approach for the calculation of winding resistance in gapped magnetic components. In Proceedings of the 2014 IEEE Applied Power Electronics Conference and Exposition, Fort Worth, TX, USA, 16–20 March 2014.
31. Massarini, A.; Kazimierczuk, M. Self-capacitance of inductors. *IEEE Trans. Power Electron.* **1997**, *12*, 671–676. [CrossRef]

© 2020 by the authors. Licensee MDPI, Basel, Switzerland. This article is an open access article distributed under the terms and conditions of the Creative Commons Attribution (CC BY) license (http://creativecommons.org/licenses/by/4.0/).

Article

Fault Investigation in Cascaded H-Bridge Multilevel Inverter through Fast Fourier Transform and Artificial Neural Network Approach

G. Kiran Kumar [1], E. Parimalasundar [2], D. Elangovan [1,*], P. Sanjeevikumar [3,*], Francesco Lannuzzo [4] and Jens Bo Holm-Nielsen [3]

1. School of Electrical Engineering, VIT Vellore, Tamil Nadu 632014, India; kiran215vit@gmail.com
2. Department of Electrical and Electronics Engineering, Sree Vidyanikethan Engineering College, Tirupati 517102, India; parimalpsg@gmail.com
3. Center for Bioenergy and Green Engineering, Department of Energy Technology, Aalborg University, 6700 Esbjerg, Denmark; jhn@et.aau.dk
4. Department of Energy Technology, Aalborg University, 9220 Aalborg, Denmark; fia@et.aau.dk
* Correspondence: elangovan.devaraj@vit.ac.in (D.E.); san@et.aau.dk (P.S.)

Received: 24 December 2019; Accepted: 7 March 2020; Published: 11 March 2020

Abstract: In recent times, multilevel inverters are used as a high priority in many sizeable industrial drive applications. However, the reliability and performance of multilevel inverters are affected by the failure of power electronic switches. In this paper, the failure of power electronic switches of multilevel inverters is identified with the help of a high-performance diagnostic system during the open switch and low condition. Experimental and simulation analysis was carried out on five levels cascaded h-bridge multilevel inverter, and its output voltage waveforms were synthesized at different switch fault cases and different modulation index parameter values. Salient frequency-domain features of the output voltage signal were extracted using a Fast Fourier Transform decomposition technique. The real-time work of the proposed fault diagnostic system was implemented through the LabVIEW software. The Offline Artificial neural network was trained using the MATLAB software, and the overall system parameters were transferred to the LabVIEW real-time system. With the proposed method, it is possible to identify the individual faulty switch of multilevel inverters successfully.

Keywords: Artificial Neural Networks (ANN); fault diagnosis; Fast Fourier Transform (FFT); Multilevel Inverter (MLI); LabVIEW

---

1. Introduction

In recent years, multilevel inverters are drawing intense interest in the research of solid industrial electric drives organizes in the direction of attaining the high power demands necessary with them. The foremost merits of Multilevel Inverters (MLIs) are minimization of harmonic deformation of the output voltage waveform by way of raising incapacity of levels as well as litheness for the usage of battery sets or fuel for in-between periods [1–3].

Although MLIs are effectively used in engineering applications employing a confirmed technology, the collapse of power electronic switches and its fault investigation is until now a recent research issue for researchers. In engineering applications, it is implemented to examine the state of power switches which is available in inverters. The extent of levels in the inverter changes, the number of power also switches varied, which can raise the chance of collapse of any one of the switches; hence, any such fault should be acknowledged at the initial stage so that the process of drive and motor during anomalous conditions is not affected [4–10]. The different modes of the collapse of power semiconductor switches, an open-switch as well as the short-switch fault, directs to current harmonics and generates troubles in the gate driving circuits. Therefore, it reduces the system's concert. Several researchers used the inverter output current and voltage for constructing that fault identification arrangement [11,12]. Surin Khomfoi et al. [13], created an open-switch fault analytic coordination of an MLI created based on that output voltage with Fast Fourier Transform (FFT) model as well as five parallel neural networks employing 40 contribution neurons for every system. While the range of that neural network is extremely complex because of 40 input neurons, inside of a new paper, Surin Khomfoi et al. projected a different method that contains a mixture of FFT principal constituent analysis, genetic algorithm as well as neural network technology for identifying the fault category as well as fault location in an inverter [14].

Recognition of faulty switches of MLIs is still an emerging research area, and numerous researchers are steadily working on the way to identify the faults precisely. On the other hand, information about the actual instance implementation of high concert fault investigation system for cascaded H-bridge multilevel inverter is inadequate. With that consideration, Multilevel Inverter (MLI) output voltages are measured as a significant constraint to identifying faulty switches. Real-time implementation of overall fault analytic schemes has been executed in National Instruments (NI)—LabVIEW software with a version of DAQmx 19.1; it has been a complicated apparatus designed for rising as well as operating actual instant submission. LabVIEW makes use of graphical encoding language formed by National Instruments and is also successfully applied for data attaining, instrumentation control, as well as in automotive industries. ANN is a useful tool in the classification of patterns in the course of learning as well as nonlinear mapping. Together, LabVIEW and ANN are arranged to measure. It is essential as an additional test for obtaining research work. Therefore, the actual instant fault detection method is programmed at some stage in LabVIEW with the help of ANN [15–17].

2. Structure of H-Bridge Multilevel Inverter

The H-Bridge inverter modules with separate DC sources are connected in series to form a cascaded multilevel inverter. Figure 1 illustrates a typical three-phase cascaded MLI using 3 H-bridge modules within every phase associated with a three-phase asynchronous motor load. The number of output voltage levels can be calculated with the formula 2S+1; here, 'S' is the number of H-Bridge modules utilized for it. The 3 phase MLI structure is generally used for industrialized applications. In current work, the single-phase MLI is used because that projected fault analytic scheme has the capability of expanding in favour of three-phase applications.

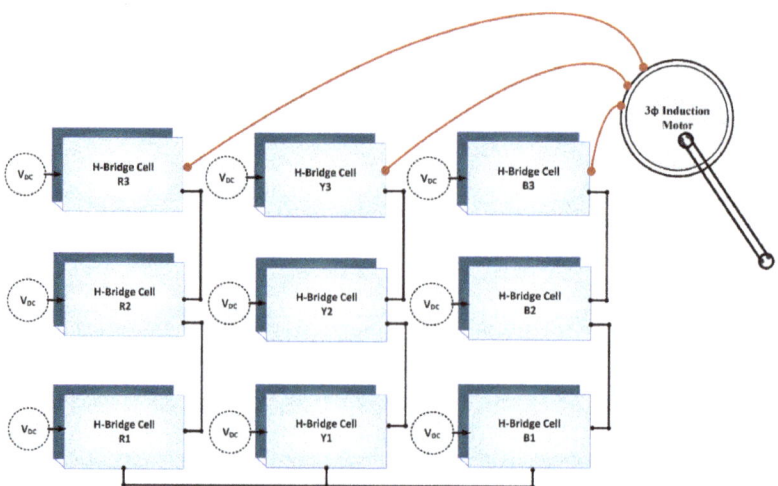

Figure 1. Representation of a three-phase H-bridge cascaded multilevel inverter fed induction motor.

Figure 2 illustrates the representation of the single-phase five levels cascaded H-bridge inverter utilized; it contains 2 H-Bridge modules, as well as 8 number of powers, switches Insulated-Gate Bipolar Transistors (IGBTs). Every IGBT switch is named following its module location like S1A, S2A, S3B, S4B, etc. The cascaded inverter has been linked along with a dynamic load like 1ψ, 0.5 HP, 50 Hz asynchronous motor. The Sinusoidal Pulse Width Modulation (SPWM) technique has been applied for generating the necessary switching signals for IGBTs. In this SPWM, higher frequency contained triangular carrier waves are used along with a sinusoidal reference wave. Figure 3 shows that production of switching sequence corresponding to module A at a carrier wave frequency value (fc) of 3 kHz as well as a modulation index value (m) of sinusoidal wave 0.85. Now in this work, the modulation index is deferred in-between sort of values 0.8 to 0.95.

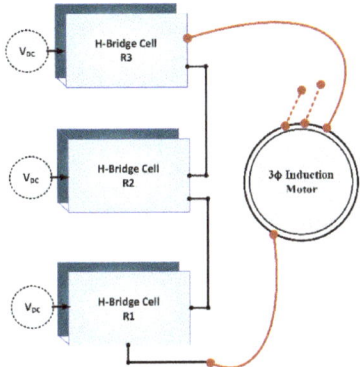

Figure 2. Simulink model of 1-Φ cascaded H-Bridge 5-level inverter fed with Induction Motor load.

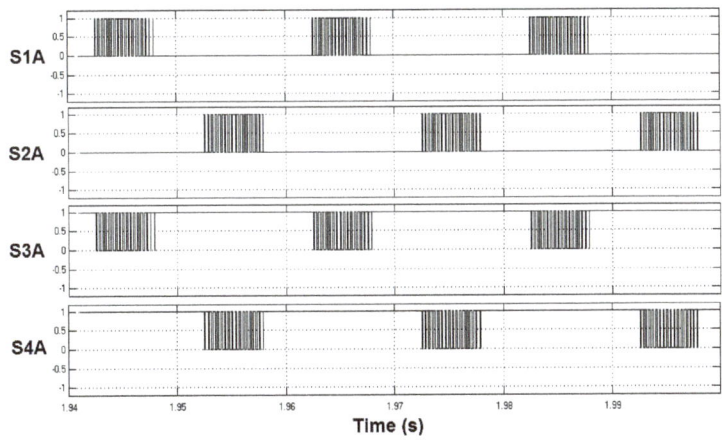

Figure 3. Sinusoidal Pulse Width Modulation cell A switching patterns created with a carrier wave frequency of 3 kHz and modulation index value of 0.85.

3. Fault Analysis of Output Voltage and Current

Simulation fault analysis was carried out with the help of Matlab/Simulink (R2019b) to realize the output load voltage, as well as output, load current waveform earlier than as well as later than the fault commencement of the MLI. Foremost, the open circuit fault occurs at 1 s on switch S1A of module A. Figure 4 shows the actual load voltage of the inverter, its exaggerated outlook as well as current wave earlier and later of that faulty condition of single-phase cascaded H-Bridge MLI associated employing the induction motor load. The voltage, as well as current patterns, were exampled for 20 kHz. Likewise, Figure 5 shows the output load voltage, as well as the output, load current waveforms later than the fault creation of open-circuit fault in switch S2A of Module A.

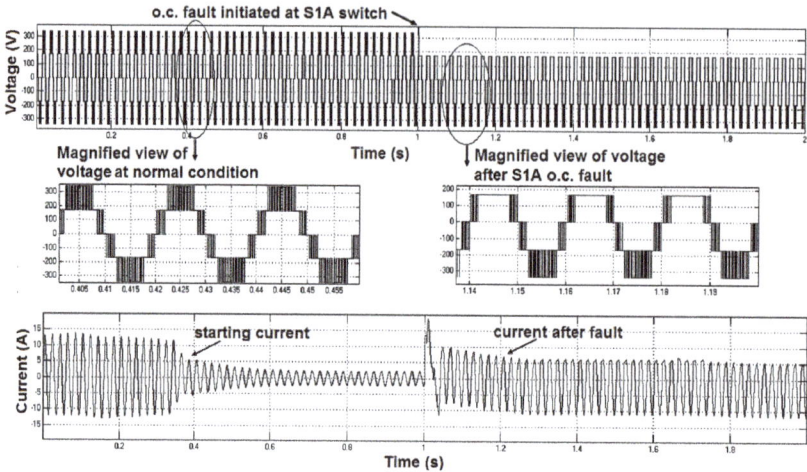

Figure 4. Load current and voltage waveform analysis of Bridge—A of S1A during an open-circuit fault.

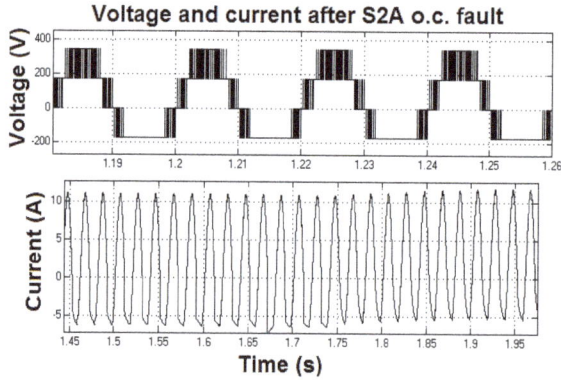

Figure 5. Load current and voltage waveform analysis of Bridge—A of S2A during an open-circuit fault.

4. Concept of Fast Fourier Transform and Feature Extraction Process

In the Fast Fourier Transform method, the output voltage waveform features are extracted. In order to generate a powerful fault analysis method, it is essential on the way to execute frequency domain analyses for the output load voltage patterns. The FFT method has been used for pulling out different parameters of the output voltage signal. As seen, signals of output parameters are not easy to be measured as a significant attribute to categorizing a faulty assumption. Consequently, the signal conversion method is required. A proper choice of that characteristic extractor is to give sufficient important information about the neural network in the example set; thus, that was the maximum amount of precision within that neural network concert that was attained. Single probable method of execution with Digital Signal Processing (DSP) microchip is executed with the help of FFT [9]. Initiating through Discrete Fourier Transform in Equation (1), after that FFT technique is with a combination of decimation within time decomposition algorithm has been represented in Equations (2) and (3):

$$F_k = \sum_{n=0}^{N-1} f_n W_N^{nk} \quad \text{for} \quad k = 0, \ldots, N-1 \qquad (1)$$

where $W_N = e^{-j\frac{2\pi}{N}}$

$$F_k = G_k + W_N^k H_k \quad \text{for} \quad k = 0, ,\frac{N}{2} - 1 \qquad (2)$$

$$F_{k+\frac{N}{2}} = G_k - W_N^k H_k \quad \text{for} \quad k = 0, ,\frac{N}{2} - 1 \qquad (3)$$

G_k is to be on behalf of even-numbered essentials of f_n, while H_k is to be on behalf of odd-number essentials of fn. G_k, Z as well as Hk, are to be evaluated as exposed in Equations (4) and (5).

$$G_k = \sum_{n=0}^{\frac{N}{2}-1} f_{2n} W_{\frac{N}{2}}^{nk} \qquad (4)$$

$$H_k = \sum_{n=0}^{\frac{N}{2}-1} f_{2n+1} W_{\frac{N}{2}}^{nk} \qquad (5)$$

Figure 6 shows the schematic diagram representation of that fault analysis system implemented on behalf of the recognition of the failures in power electronic switches using those features extracted from the FFT technique. The inverter load voltage wave has been deliberated and using FFT technique, essential functions of the voltage signal, i.e., Total Harmonic Distortion (THD) (%), harmonic/fundamental ratio (%) values up to 11th harmonics are extracted. Hence, the RMS value analysis is also simultaneously carried out in the output voltage signal, and the extracted 12 features are specified as input for that the Artificial Neural Network. Figure 6 illustrates the feature 100/50 representing the ratio of the second harmonic to fundamental; feature 150/50 represents the ratio of third harmonic/fundamental and so on. The output of the trained patterns of the ANN with LabVIEW software identifies the faulty switch of the multilevel inverter. Therefore, the FFT method is useful for every load voltage waveform underneath the open circuit fault condition as well as short circuit fault conditions with corresponding harmonic level variations, Vrms, and THD values. Figure 7 illustrates the FFT frequency plot of output voltage during a healthy and open-circuit fault condition. Figure 8 illustrates different features of the load voltage wave obtained from the FFT technique on various open switch faulty cases from second harmonic ratios to 11th Harmonic ratios. Figure 9 illustrates the features of the load voltage wave obtained from the FFT technique on various short switch faulty cases for different harmonic ratio values. Figure 10 shows THD of the output voltage waveform obtained from the FFT technique on various faulty cases as well as on various values of modulation index at different switches. Figure 11 illustrates the RMS values of the output voltage waveform obtained on various faulty cases as well as on various values of modulation index at different switches. Extracting the distinct features from the waveform of load voltage will automate the fault analysis process.

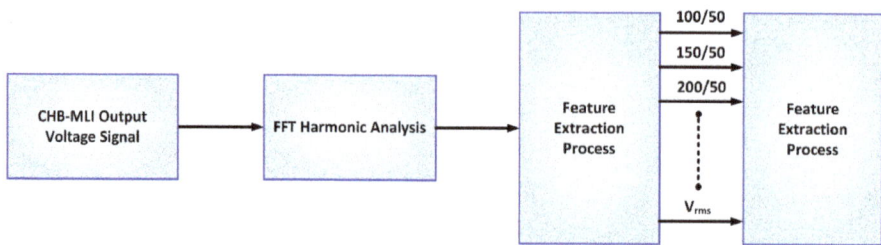

Figure 6. LabVIEW based fault investigative system using FFT features.

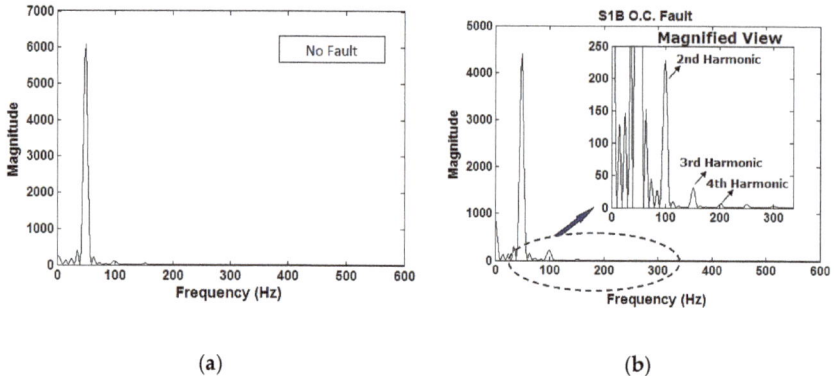

Figure 7. Frequency plot of the output voltage. (**a**) Normal condition; (**b**) S1B open circuit fault.

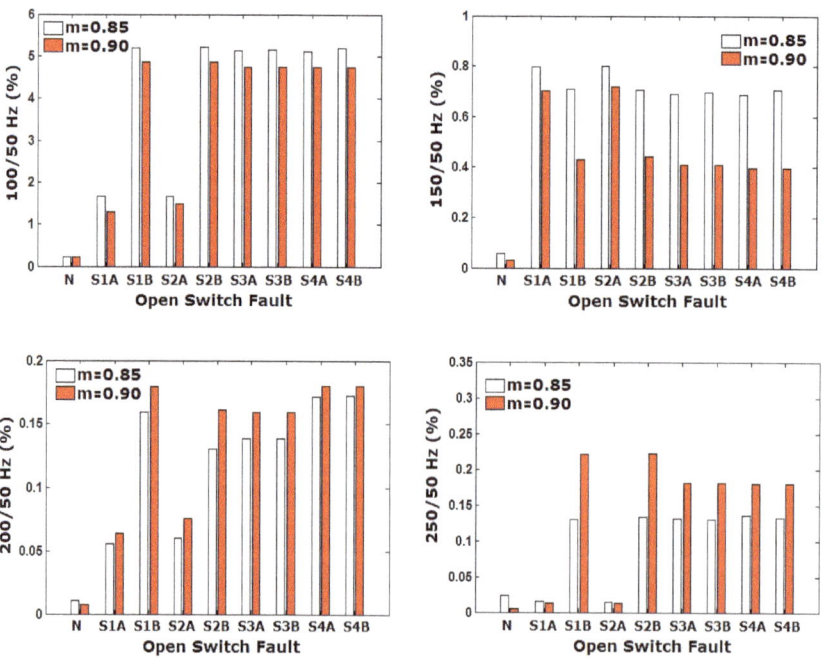

Figure 8. *Cont.*

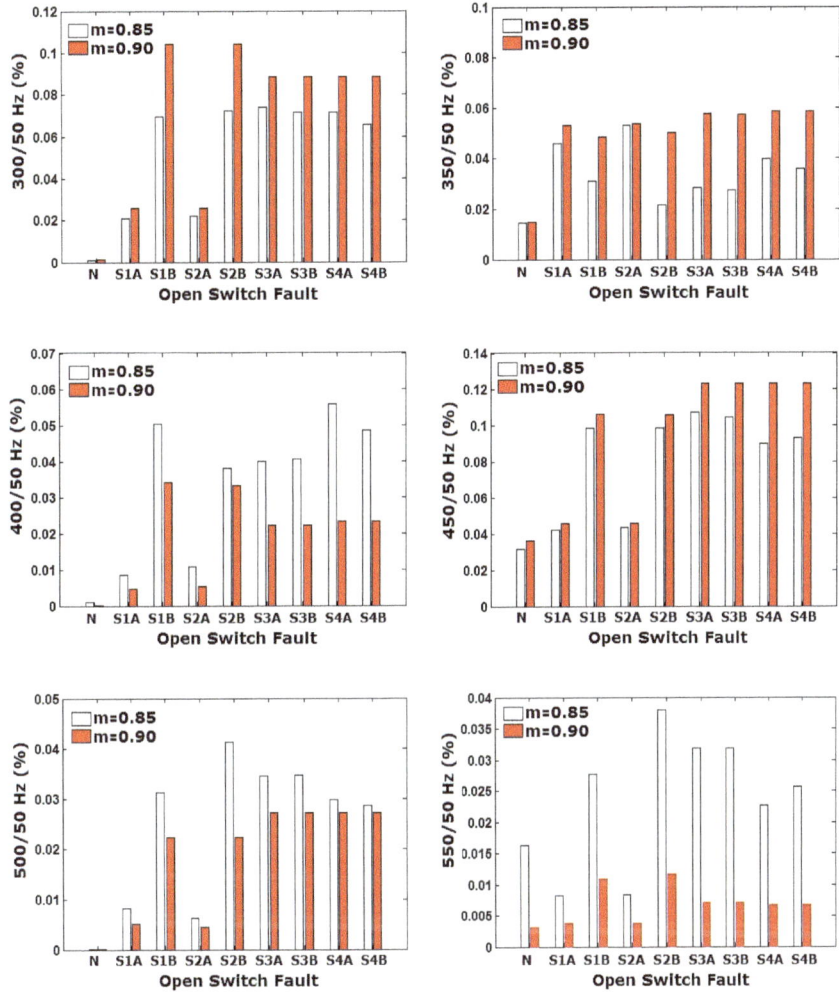

Figure 8. Features of the output voltage waveform acquired from FFT technique at distinct open switch fault cases from 2nd harmonic ratios to 11th harmonic ratios.

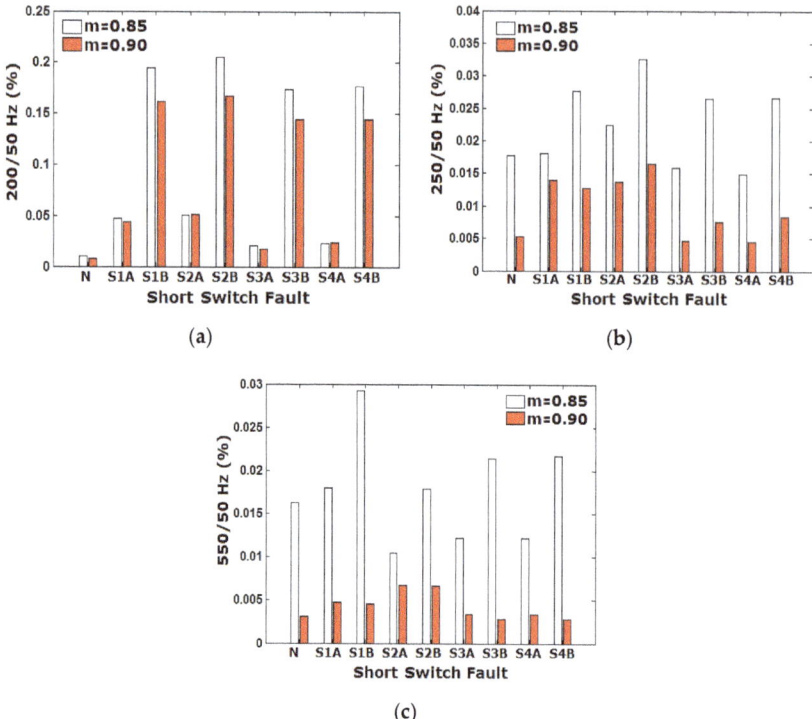

Figure 9. Features of the voltage waveform acquired from the FFT technique at distinct short switch fault cases (**a**) 4th harmonic ratio, (**b**) 5th harmonic ratio, (**c**) 11th harmonic ratio.

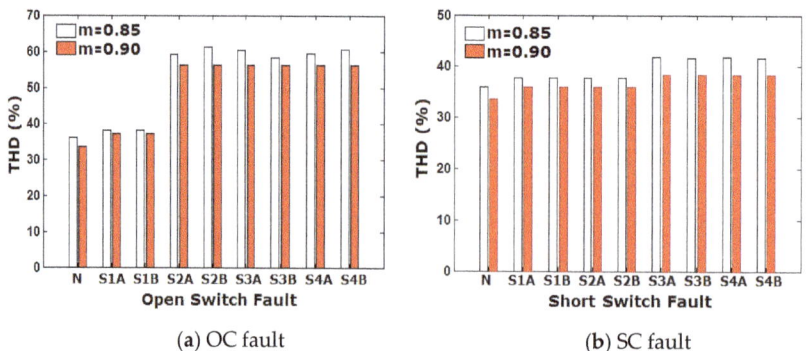

(a) OC fault (b) SC fault

Figure 10. THD value of the output voltage waveform obtained from the FFT technique at different fault cases and different modulation index values.

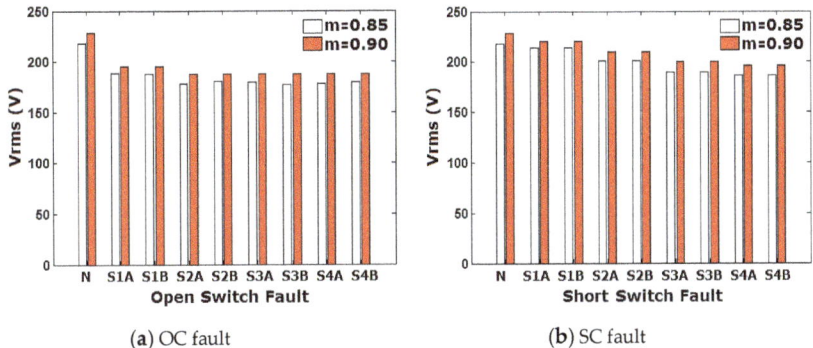

(a) OC fault   (b) SC fault

Figure 11. Output voltage waveforms attained at distinct modulation index values at various fault conditions.

5. Structure of Fault Diagnostic System

Figure 12 shows the schematic diagram of the overall fault diagnosis scheme built to recognize the power electronic switch failure in MLI. The hardware arrangement contains a DC Source, MLI, Induction Motor, as well as A-Data Fetching System.

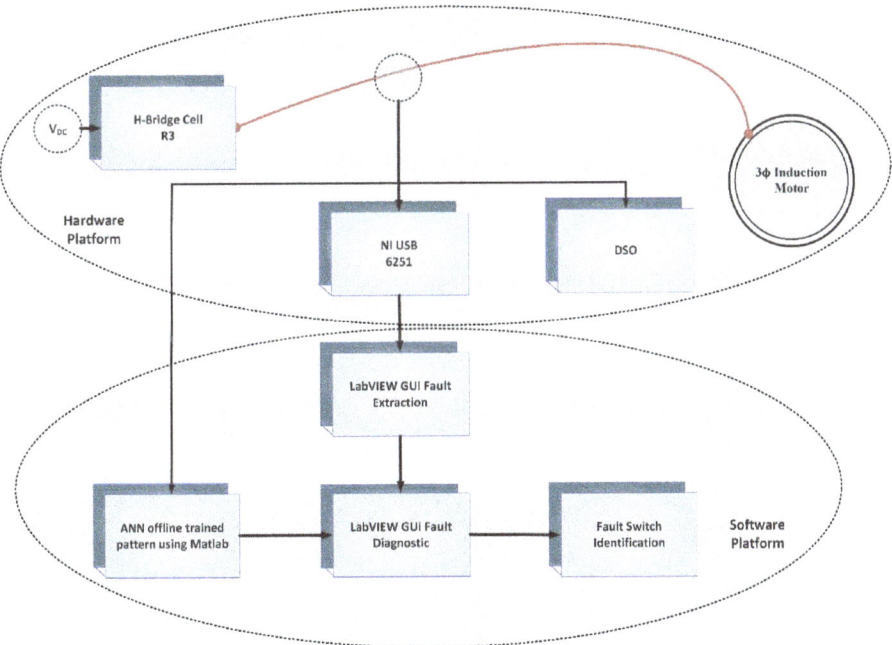

Figure 12. Implementation of the overall fault analytical system.

Primarily, extracting the voltage wave features as well as various modulation index values using FFT harmonic analysis in Matlab/Simulink (R2019b), the various fault conditions have been created in both practical setting and simulation. The obtained features were given to ANN offline training with Matlab. This trained model contains both weight values, as well as bias values of ANN, are fed on the way to graphical programming language LabVIEW used for fault analysis. For actual instant submission, outputs of voltage sensors are specified toward NI Universal Serial Bus data acquisition system that has been attached with PC. Voltage information is operated in LabVIEW FFT attribute, taking out an examination as well as compared by way of that offline trained example. After that, the LabVIEW GUI indicates fault switches in the MLI that helps maintain the system's reliability.

Figure 13 shows the laboratory experiment arrangement worn to gather load voltage waveforms of MLI on behalf of various switch faulty situation. To obtain five-level output voltages, the cascaded arrangement of PWM modules triggered 2 H-bridge inverters must be created. IGBTs having the specifications of 600 V, 25 A are chosen as switches. PWM module containing the reference as well as carrier signal selecting, modulation index as well as switching frequency modification are worn for sending necessary gate signals to IGBTs.

Figure 13. Laboratory experimental arrangement.

Figure 14a,b show the data acquisition system interfacing to LabVIEW software in the computer on behalf of actual instant submissions. National Instruments (NI) USB-6251 (1.25 MSa/Sec) are worn as data fetching arrangement that are interfaced with the computer to record as well as facilitate the new process for obtaining signals.

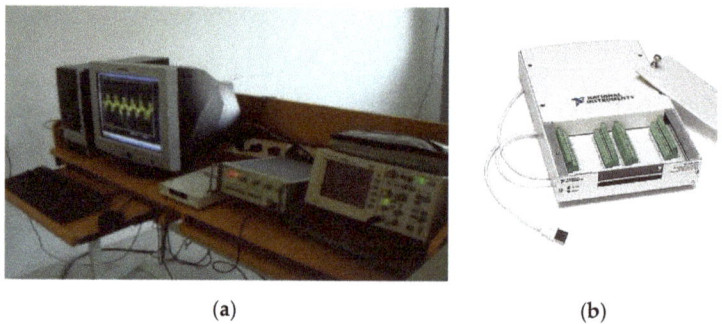

(a)   (b)

Figure 14. (a) Data acquisition system interfaced with LabVIEW software in PC (b) NI USB-6251 hardware.

The total arrangement can measure the 16 analogue signals as well as 16 bits. The Digital Storage Oscilloscope (DSO) and Agilent (1 GSa/Sec), are worn as well for visualizing load voltage waves. Open circuit and short circuit faulty conditions are formed on every switching device, and consequent output voltage waves are stored. A voltage sensor has been utilized for collecting the signals at the output and is directly interfaced to NI USB-6251. The output signals are extracted from inverter for variety of indexed modularity values, and FFT harmonic analysis is also conceded. A significant characteristic of that voltage sensor is to extract the energy substance of that signal at various levels of dissolution.

5.1. Simulation Results at open circuit Fault

Firstly, an open circuit faulty condition is subjected to switch S1A, and then the resulting load voltage waveform recorded. Likewise, open circuit faulty condition is formed on the remaining switches of H-Bridge A as well as H-Bridge B. The resulting output voltage waveforms are recorded for an advanced characteristic of pulling out progression. Figure 15 shows the distinctive output waveforms attained when open circuit switch faulty condition on H-Bridge A. For assessment, that output voltage waveform when in the no-fault situation has also been included in Figure 15. These output voltage waveform patterns illustrate to find variations between healthy as well as faulty conditions effortlessly.

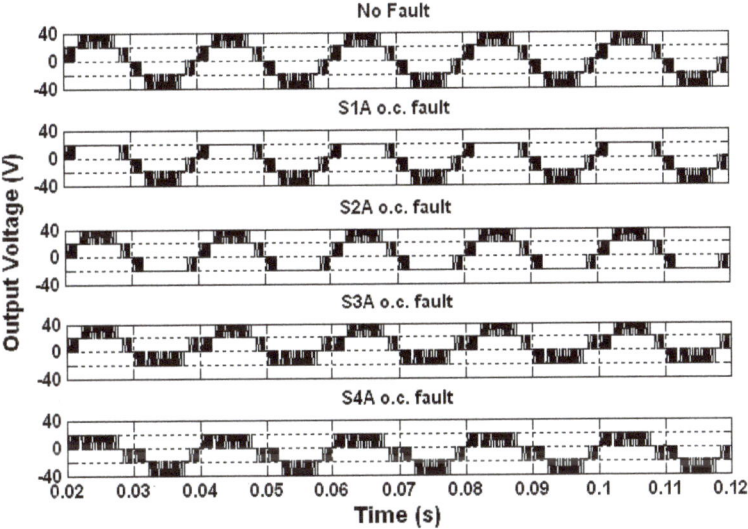

Figure 15. Fault condition output voltage waveforms.

5.2. Simulation Results at short circuit Fault

During this condition, short circuit fault has been formed on every one switch of H-Bridge A and H-Bridge B, one after one as well as that resulting output voltage waveforms are also recorded on behalf of supplementary characteristic pulling out method. Figure 16 shows that characteristic output voltage waves attained when normal conditions as well as short- switch faulty circumstances of H-Bridge B. by observing that the countable dissimilarity has been there in every output voltage waves of short circuit unsatisfactory situation while comparing with the reasonable condition. FFT method has been used for every output voltage wave underneath short circuit fault circumstance, and then consequent FFT harmonic analysis is evaluated.

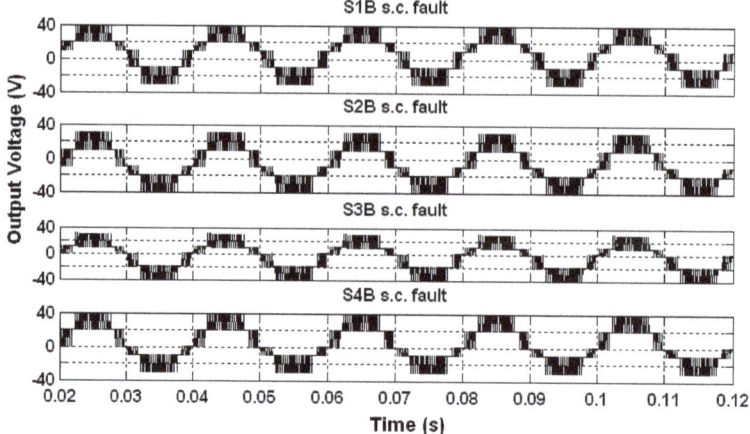

Figure 16. Short Circuit fault condition output voltage waveforms.

6. Experimental Validation

6.1. Feature Extraction Analysis Using LabVIEW

The whole faulty diagnosis system has implemented with National Instruments (NI) - LabVIEW software with a version of DAQmx 19.1. Figure 17 shows the implemented LabVIEW frontage panels containing the output voltage imprison as well as examination unit before identifying the fault switch. It shows the initial Data Acquisition capture settings in terms of max and min value, sampling frequency, number of samples per channel, multi-factor value, the magnitude of RMS voltage in max and min and time scale parameters. The GUI improved within LabVIEW displayed the fetched output waveform of voltage have frontage on the side panel of that programming algorithm. Their front side panel behaves similar to a UI someplace the user be able to set up as well as pull out information. Formerly that NI USB gadget has correctly interfacing to that LabVIEW front side panel; the piece of equipment contact indicates in green colour blinking. The front side panel has had power over parameters of output signal like the scale of time as well as magnitude, sampling frequency as well as no. of samples of every indication.

Moreover, with the help of acquisition setting the no. of sample signal capturing is also controlled for a particular time. Separate sub VI has power over frequency domain investigation of that output voltage signal. Because of the NI Data Acquisition System (DAS) as well as LabVIEW software's has the capability to capturing as well as analyzing the data set on a precise instant. Signals are capturing endlessly as well as recorded in the computer on behalf of supplementary dispensation. This front panel also shows the variations in the RMS value of the output voltage signal concerning time which helps in understanding the trend analysis of the Vrms parameter. Figure 18 shows that front end side panel of LabVIEW intended to examination frequency domain of the output voltage signals using FFT technique. This unit contains the have power over choosing the signals for FFT analysis purpose. Within this unit, control also provided for tracking of individual FFT plot of the voltage signal. Peak values of the harmonic frequencies are used to evaluate the harmonic ratios concerning fundamental frequencies. The Fast Fourier Transform frequency domain harmonic examination of the front panel is improved within the research work as well as it shows in Figure 19. This has been deliberated on the way toward the observation of various harmonic/fundamental ratios, and THD value of the output voltage signal of multilevel inverter and Figure 20 shows that VI front panel of MLI faulty switch analysis. In that screen, that is shown the possibility of tracking the harmonic scheme for each separate output voltage signals. This software module is developed in such a way to evaluate up to

11th harmonic ratio. Trend analysis of THD value and harmonic ratios is possible in this front panel and the faulty.

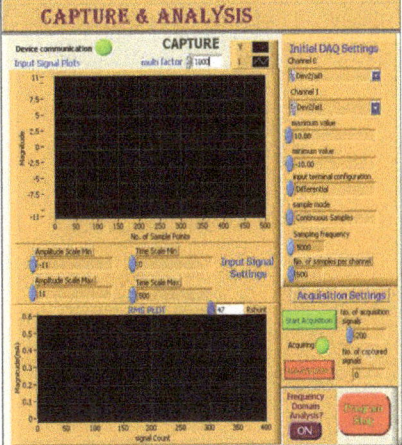

Figure 17. LabVIEW output Voltage imprisons in the front panel and analysis of multilevel inverter.

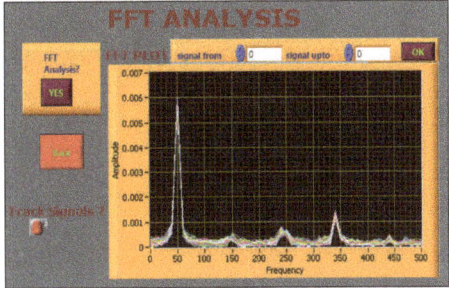

Figure 18. LabVIEW FFT based frequency domain analysis in the front panel.

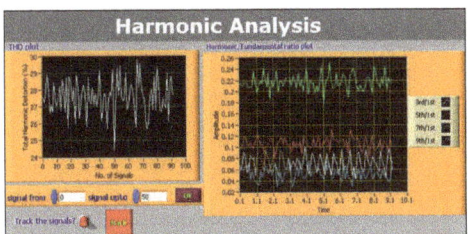

Figure 19. The front panel of FFT based THD and Harmonic analysis.

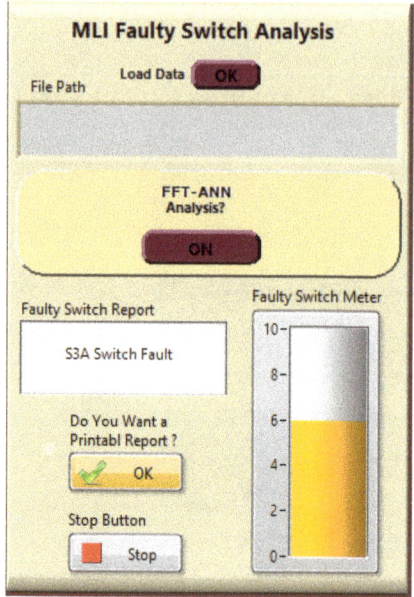

Figure 20. The front panel of MLI faulty switch analysis.

The switch must be identified and Figure 21 illustrates the output voltage pattern that relates to MLI under various faulty switch conditions at real-time implementation.

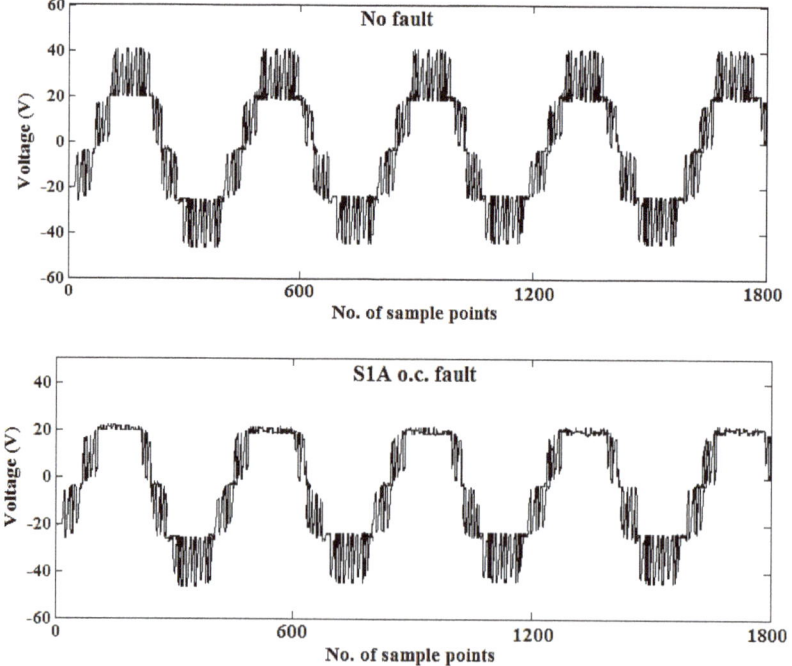

Figure 21. *Cont.*

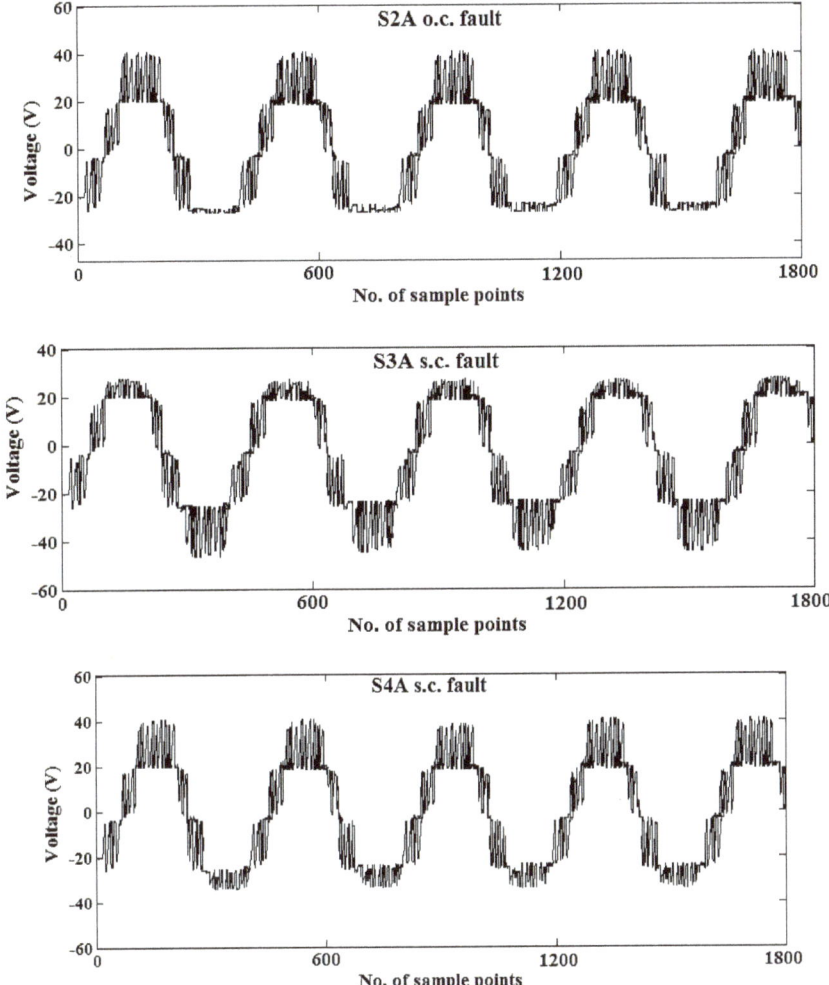

Figure 21. Voltage waveforms acquired at different open circuit and short circuit switch fault situations.

6.2. Real-Time Fault Diagnosis Results from LabVIEW-ANN Approach

On the way towards the mechanize the development that of fault analysis in MLI, a multilayer feed-forward network, as well as a backpropagation learning algorithm was utilized [15].

Figure 22 shows the ANN schematic diagram. It has a structure containing an input layer, one forbidden layer, and one output layer. The targeted and input vectors are primarily fed with some values for a training network. Exercising the arrangement is completed by altering the weight as well as the bias of the unit depends among the significant fault. Backpropagation training algorithm contains a frontward pass as well as toward the back pass is conceded out in anticipation of the Mean Square Error (MSE) has been evaluated up to the lowest value. The collected data is achieved at what time error between them calculated as well as the preferred amount produced, which is a lesser amount of that set value.

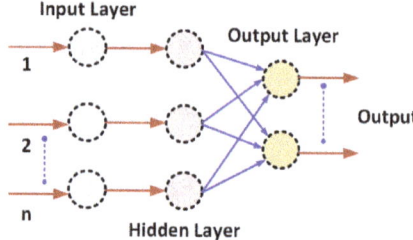

Figure 22. Schematic diagram of an Artificial Neural Network.

Here MSE, that has been meaning of the number of errors for every set of input as well as resulting output, been evaluated with Equation (6):

$$MSE = \frac{1}{m}\sum_{k}^{m}(S_k - Y_k)^2 \qquad (6)$$

Here S_k, as well as Y_k are, correspondingly, the preferred, as well as measured output on behalf of kth input set and m, is the whole quantity of their output parameters [15]. Here information of the revised neural network is utilized and is displayed in Table 1.

Table 1. Specifications of FFT—ANN-Lab VIEW Approach.

No. of Inputs	12
No. of Neurons in Hidden Layer	24
No. of Neurons in Output Layer	9
Learning Rate (η)	0.1
No. of Iterations	3800
No. of Training Sets	200
No. of Test Input Sets	150
Convergence Criteria	0.01

Within the proposed work, 12 parameters (10 harmonic ratios, THD and Vrms) obtained as features from the FFT technique of a faulty conditioned output voltage signals are fed to the input of the neural network. There is a total of 9 produced neurons for classifying that fault as that of no-fault, S1A fault up to S4B fault are shown in Table 2.

Table 2. Training pattern of Neural Network.

Classification of Fault	Position of Neuron	Output Pattern
No fault	1	[1 0 0 0 0 0 0 0 0]
S1A fault	2	[0 1 0 0 0 0 0 0 0]
S1B fault	3	[0 0 1 0 0 0 0 0 0]
S2A fault	4	[0 0 0 1 0 0 0 0 0]
S2B fault	5	[0 0 0 0 1 0 0 0 0]
S3A fault	6	[0 0 0 0 0 1 0 0 0]
S3B fault	7	[0 0 0 0 0 0 1 0 0]
S4A fault	8	[0 0 0 0 0 0 0 1 0]
S4B fault	9	[0 0 0 0 0 0 0 0 1]

It is shown that the ANN training sequence approaches various switches faults in MLI. In the exercise sequence, each neuron in that output layer of the neural network is assigned to particular faults and then trained for a binary value of 1 or 0, as shown in Table 2. For example, in the case of the no-fault condition, the first neuron in the output layer has been assigned a value of 1, and all other neurons are trained for a value of 0. Similarly, for different fault cases, the output layer neurons are trained for different binary training patterns. For the offline training of the neural network, 200 training sets were used, and the weight matrix of the trained pattern is given as an input to the LabVIEW GUI module for testing purposes with 150 test inputs.

Figure 23 shows the performance of the network at various iterations. The training of the current system reaches the junction criterion after close to 3800 iterations. It is proved that 3800 iterations were enough to know the successful training of that revised neural network. Consequently, a concert of that back PNN is known by way of 24 forbidden layer neurons maintained that worth of that learning rate is 0.1, and the number of iterations is 3800.

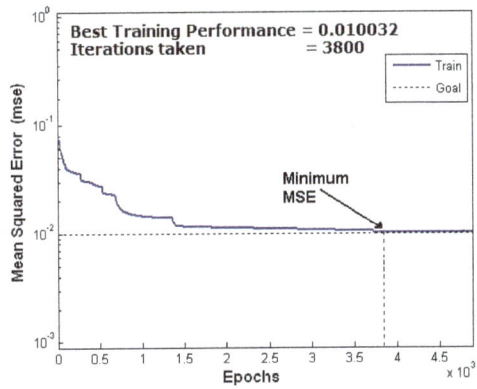

Figure 23. Variations occurred in MSE of the ANN during training pattern concerning an increase in the number of iterations.

In general, it is noticed that the neural network accurately predicts the no-fault case at all tested numbers of forbidden layer neurons. Since that network convergence has not been reached within the specified revised neural network parameters in such cases of 15 or 20 forbidden layer neurons, then the accuracy of identification tempo is affected. It has identified their concert of the neural network been enhanced on behalf of 24 forbidden layer neurons while comparing employing further situations. The average recognition tempo on behalf of every faulty case is 100% in the considered case, as well as the neural network, has capable of discovering that fault in all faulty cases effectively. Table 3 gives a detailed analysis of the identification rate, and Figure 24 illustrates the evaluation of neural network means the square error of various numbers of forbidden layer neurons.

Table 3. Overall Identification rates of FFT-ANN-LabVIEW approach.

Classification of Fault	Identification Rate (%) at Different Number of Hidden Layer Neurons		
	15	20	24
No fault	100	100	100
S1A fault	91	92	100
S1B fault	93	95	100
S2A fault	92	95	100
S2B fault	91	92	100
S3A fault	95	95	100
S3B fault	92	96	100
S4A fault	93	95	100
S4B fault	92	94	100

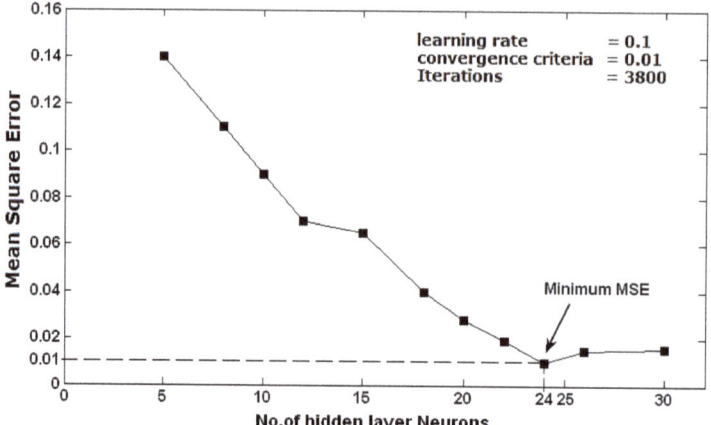

Figure 24. Evaluation of the mean square error of the neural network at different numbers of hidden layer neurons.

Real-time implementation of LabVIEW is dependent on the trail of the fault analysis of MLI with load voltage characteristics like FFT harmonic analysis showing the possibilities of determining the switch failure in a particular position in an MLI. When comparing with other techniques mentioned in previous publications [13,14], the projected technique considerably decreases the number of inputs to ANN network as well as analyzes the occurrence of faulty conditions in 10 msec, and it can find the faulty condition of an exact switch (OC fault or SC fault) in the MLI. Also, the projected method gives a 100% detection rate among no-fault as well as fault conditions of a switch. Therefore, on one occasion, the switch fault is analyzed that has helped operate and carry out precautionary safeguarding.

7. Conclusions

This paper's proposed H-bridge cascaded five level multilevel inverters associated through a load of induction motor with faulty switch analysis is carried out. First, the significant attributes of load voltage output waveform are examined with simulation as well as experimental analysis on dissimilar open-switch and short-switch faulty conditions. Meanwhile, the existing problems are further examined. Furthermore, essential features like harmonic analysis with FFT technique are specified to be input to that backpropagation trained ANN along with an evaluation of the mean square error at the different number of hidden layer neurons; to make the neural network offline the Matlab software is utilized. The actual instance function of that anticipated fault diagnosis scheme is executed with the LabVIEW software. This projected fault diagnosis scheme has the potential to accurately recognize each separate fault switch of the cascaded multilevel inverter. It can categorize 100% precisely typical as well as fault situations. Therefore, in this instance, the faulty switch conditions were recognized immediately, and implementing this system will help the operator in performing protective maintaining work.

Author Contributions: G.K.K., and E.P., have developed the proposed research concept, and they both are involved in studying the execution and implementation with statistical software by collecting information from the real environment and developed the simulation model for the same. D.E., P.S., F.L., J.B.H.-N. shared their expertise and validation examinations to confirm the concept theoretically with the obtained numerical results for its validation of the proposal. All authors are to frame the final version of the manuscript as a full. Moreover, all authors involved in validating and to make the article error-free technical outcome for the set investigation work. All authors contributed to the research investigation equally and presented in the current version of the full article. All authors have read and agreed to the published version of the manuscript.

Funding: No source of funding for this research activities.

Conflicts of Interest: The authors declare no conflict of interest.

References

1. Zheng, Z.; Wang, K.; Xu, L.; Li, Y. A hybrid cascaded multilevel converter for battery energy management applied in electric vehicles. *IEEE Trans. Power Electron.* **2014**, *29*, 3537–3546. [CrossRef]
2. Javad, G.; Reza, N. Analysis of Cascaded H-Bridge Multilevel Inverter in DTC-SVM Induction Motor Drive for FCEV. *J. Electr. Eng. Technol.* **2013**, *8*, 304–315.
3. Banaei, M.R.; Salary, E. A New Family of Cascaded Transformer Six Switch Sub-Multilevel Inverter with Several Advantages. *J. Electr. Eng. Technol.* **2013**, *8*, 1078–1085. [CrossRef]
4. Ui-Min, C.; Lee, K.-B.; Frede, B. Diagnosis and tolerant strategy of an open-switch fault for T-type three-level inverter systems. *IEEE Trans. Ind. Appl.* **2014**, *50*, 495–508. [CrossRef]
5. Chen, A.; Hu, L.; Chen, L.; Deng, Y.; He, X. A multilevel converter topology with fault-tolerant ability. *IEEE Trans. Power Electron.* **2005**, *20*, 405–415. [CrossRef]
6. Pablo, L.; Josep, P.A.; Thierry, M.; Jose, R.; Salvador, C.; Frédéric, R. Survey on fault operation on multilevel inverter. *IEEE Trans. Ind. Electron.* **2010**, *57*, 2207–2218.
7. Ma, M.; Hu, L.; Chen, A.; He, X. Reconfiguration of carrier-based modulation strategy for fault-tolerant multilevel inverters. *IEEE Trans. Power Electron.* **2007**, *22*, 2050–2060. [CrossRef]
8. Diallo, D.; Benbouzid, M.H.; Hamad, D.; Pierre, X. Fault detection and diagnosis in an induction machine drive—A pattern recognition approach based on concordia stator mean current vector. *IEEE Trans. Energy Conv.* **2005**, *20*, 512–519. [CrossRef]
9. Estima, J.; Cardoso, A.M. A new algorithm for real-time multiple open-circuit fault diagnosis in voltage-fed PWM motor drives by the reference current errors. *IEEE Trans. Ind. Electron.* **2013**, *60*, 3496–3505. [CrossRef]
10. Khan, M.A.S.K.; Rahman, M.A. Development and implementation of a novel fault diagnostic and protection technique for IPM motor drives. *IEEE Trans. Ind. Electron.* **2009**, *56*, 85–92. [CrossRef]
11. Lezana, P.; Aguilera, R.; Rodriguez, J. Fault detection on multicell converter based on output voltage frequency analysis. *IEEE Trans. Ind. Electron.* **2009**, *56*, 2275–2283. [CrossRef]
12. Masrur, M.A.; Chen, Z.; Murphey, Y. Intelligent diagnosis of open and short circuit faults in electric drive inverters for real-time applications. *IET Power Electron.* **2010**, *3*, 279–291. [CrossRef]
13. Surin Khomfoi, S.; Tolbert, L.M. Fault diagnostic system for a multilevel inverter using a neural network. *IEEE Trans. Power Electron.* **2007**, *22*, 1062–1069. [CrossRef]
14. Surin Khomfoi, S.; Tolbert, L.M. Fault diagnosis and reconfiguration for multilevel inverter drive using AI-based techniques. *IEEE Trans. Ind. Electron.* **2007**, *54*, 2954–2968. [CrossRef]
15. Sivakumar, M.; Parvathi, R.M.S. Diagnostic Study of Short-Switch Fault of Cascaded H-Bridge Multilevel Inverter using Discrete Wavelet Transform and Neural Networks. *Int. J. Appl. Eng. Res.* **2014**, *9*, 10087–10106.
16. Hochgraf, C.; Lasseter, R.; Divan, D.; Lipo, T.A. Comparison of multilevel inverters for static VAR compensation. In Proceedings of the Conference Record of the IEEE Industry Application Society Annual Meeting, Denver, CO, USA, 2–6 October 1994; pp. 921–928.
17. Kastha, D.K.; Bose, B.K. Investigation of fault modes of voltage-fed inverter system for induction motor drive. *IEEE Trans. Ind. Appl.* **1994**, *30*, 1028–1038. [CrossRef]

© 2020 by the authors. Licensee MDPI, Basel, Switzerland. This article is an open access article distributed under the terms and conditions of the Creative Commons Attribution (CC BY) license (http://creativecommons.org/licenses/by/4.0/).

Article

SiC-Based High Efficiency High Isolation Dual Active Bridge Converter for a Power Electronic Transformer

Mariam Saeed *, María R. Rogina, Alberto Rodríguez, Manuel Arias and Fernando Briz

Department of Electrical Engineering, University of Oviedo, 33204 Asturias, Spain; rodriguezrmaria@uniovi.es (M.R.R.); rodriguezalberto@uniovi.es (A.R.); ariasmanuel@uniovi.es (M.A.); fernando@isa.uniovi.es (F.B.)
* Correspondence: saeedmariam@uniovi.es

Received: 30 January 2020; Accepted: 1 March 2020; Published: 5 March 2020

Abstract: This paper discusses the benefits of using silicon carbide (SiC) devices in a three-stage modular power electronic transformer. According to the requirements to be fulfilled by each stage, the second one (the DC/DC isolation converter) presents the most estimable improvements to be gained from the use of SiC devices. Therefore, this paper is focused on this second stage, implemented with a SiC-based dual active bridge. Selection of the SiC devices is detailed tackling the efficiency improvement which can be obtained when they are co-packed with SiC antiparallel Schottky diodes in addition to their intrinsic body diode. This efficiency improvement is dependent on the dual active bridge operation point. Hence, a simple device loss model is presented to assess the efficiency improvement and understand the reasons for this dependence. Experimental results from a 5-kW Dual Active Bridge prototype have been obtained to validate the model. The dual active bridge converter is also tested as part of the full PET module operating at rated power.

Keywords: SiC devices; antiparallel diode; dual active bridge; power electronic transformer; high-frequency transformer

1. Introduction

A line-frequency transformer (LFT) is a key element in transmission and distribution for traditional centralized generation-based systems. Their main functionality is to interface different voltage levels in the grid [1]. LFTs are a well-established, relatively cheap, and reliable technology. However, they fail to cope with modern grid demands, such as the integration of distributed resources and energy storage systems, as well as power flow control.

The power electronic transformer (PET), also called a solid-state transformer (SST), was introduced in 1970 [2]. It is considered an alternative to LFT, as it connects two AC voltage ports while providing galvanic isolation [1]. PET is a semiconductor-based energy conversion system based on fast-switching devices, which potentially enables a significant reduction in volume and weight [3]. Moreover, thanks to the controllability of the power devices, the PET provides additional functionalities, such as reactive power, harmonics and imbalances compensation, ride-through capabilities, and smart protections.

The power semiconductor technology used in PETs has been traditionally based on Silicon (Si). However, the fast advances in wide-band-gap (WBG), specially the Silicon Carbide (SiC), power semiconductors has attracted the attention to their use in the medium voltage (MV) modular three-stage PETs [4], mainly due to their high blocking voltage along with their superior switching behavior [5,6]. Several examples of using SiC devices in different PET topologies exist in literature [7–13]. In the majority of applications, 1.2/1.7 kV commercial SiC MOSFETs are used for LV side devices, while for HV side, 10 kV non-commercial MOSFETs are used [6,8]. In some works, SiC was used in all the PET stages, such as in the TIPS (transformer-less intelligent power substation), which is an all-SiC three-stage PET topology [12]. In [14], a detailed comparison was carried out between two cases:

(1) using 10 kV and 3.3 kV SiC MOSFETs and, (2) using 10 kV and 3.3 kV Si IGBTs, for different PET topologies. Authors concluded the importance of further research into the combined use of Si IGBT with SiC MOSFETs for three-stage PETs. Moreover, authors hinted at the importance of analyzing the relevance of including a SiC antiparallel diode to justify its additional cost implication.

Accordingly, the contribution of this paper is thought in two aspects:

1. Studying the requirements of the devices in each of the PET stages to assess the benefit of integrating SiC and, consequently, identifying the stages making best use of it.
2. Selection of the SiC devices for the second PET stage, based on a dual active bridge converter (DAB). As will be explained, this stage presents the most benefits from using SiC devices. Special attention will be paid to the device operation intrinsic to the converter topology. The use of SiC MOSFETs co-packaged with an additional SiC antiparallel diode is also investigated. A loss model is developed to estimate the potential efficiency improvement in this case. Both devices (i.e., with only intrinsic body diode and with additional SiC diode) are experimentally compared at rated power to validate the proposed model.

The paper is organized as follows: Section 2 describes the selected PET topology. Section 3 studies the integration of SiC devices in the different stages of the PET and identifies the practical limitations to the enhancements introduced by the use of these devices. Section 4 discusses the DAB converter development where the selection of the devices is detailed and the improvements introduced by using a SiC antiparallel diode is studied. Experimental results are provided in Section 5.

2. PET Topology

The selected PET topology is shown in Figure 1, and was previously discussed in [15–20]. It is a three-stage modular PET, with a Cascaded H-Bridge (CHB) converter acting as the front end AC/DC converter providing a HVAC link (V_{acHV}). Each CHB cell integrates a DAB converter to provide the isolation between the HV and LV sides. Moreover, the integration of the DAB to the CHB cell provides the capability of bidirectional power transfer at the cell level [18,19]. The LV outputs of all DABs are connected in parallel to provide a high-current LVDC link (V_{dcLV}), which is connected to a three-phase four-leg (3P4L) converter providing the LVAC link (V_{acLV}).

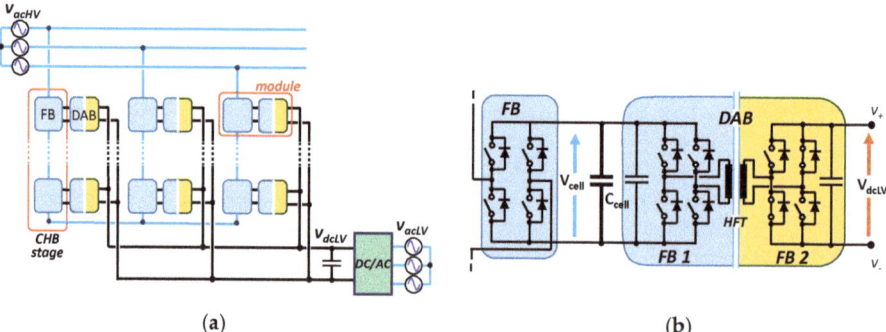

Figure 1. Selected Power Electronic Transformer (PET) structure: (a) Three-stage modular PET topology based on a Cascaded H-Bridge (CHB) converter [17]. (b) One PET module composed of a Dual Active Bridge (DAB) and the full bridge of the CHB.

As is clear in Figure 1, the structure is fully modular as it is formed by several identically-stacked cells. Thanks to this modular structure, the number of cells of the CHB is chosen in such a way that the cell voltage (V_{cell}) is equal to the required V_{dcLV}. In this way the design of this DAB is, to some extent, simplified using a unity transformation ratio (i.e., no step-up/down).

This PET topology requires high galvanic isolation between HV and LV sides. The isolation is provided by the DAB high frequency transformer (HFT). Table 1 shows the characteristics of the developed PET.

Table 1. Main Power Electronic Transformer (PET) characteristics.

Element	Parameter	Value
PET	Rated power	105 kW
	Number of cells	21 (7 per phase)
	HV/LV grid voltage (L-L)	6 kV/400 V
DAB	Rated power	5 kW
	Switching frequency (f_{sw})	30 kHz
	HFT isolation	24 kV
	Input voltage ($V_i = V_{cell}$)	800 V
	Transformer turns ratio (n)	1:1
	Leakage inductance (L_{lk})	423 µH
CHB	Rated power	5 kW
	C_{cell}	600 µF(film)

3. Use of WBG for PET

The superior material properties of WBG semiconductors allow power devices to operate at higher temperatures, voltages and switching frequency in comparison to Si counterparts [4,21]. Among WBG materials, SiC presents the most mature technology for high voltage devices [4]. Both 1.2 kV and 1.7 kV SiC MOSFETs are already available on the market with a wide range of current ratings [5,22,23]. The use of SiC MOSFETs not only introduces a relevant improvement to the efficiency of fast switching power converters, but also enables going to higher switching frequencies at high blocking voltage which cannot be achieved using available Si IGBTs. However, at limited switching frequency requirements (<10 kHz), especially for high power (>100 A), Si IGBTs are still the preferred choice due to cost-effectivity and reliability in addition to SiC MOSFET higher dv/dt, di/dt, and EMI issues [24].

Accordingly, to analyze the merits of integrating SiC devices in the PET, the device requirements for each of the three stages of the PET are identified below.

3.1. Device Requirements per PET Stage

In the addressed PET topology, since no step-up/down is needed, the power devices employed on both transformer sides (i.e., HV and LV sides) have the same voltage rating. These include devices of the CHB full bridge (FB), the two DAB full bridges (FB1, FB2), and the DC/AC converter devices (see Figure 1). While this is true, the specifications in terms of current ratings and commutation requirements differ significantly.

CHB: its power devices do not need to commutate fast due to the multilevel nature of the topology [25,26]. Moreover, C_{cell} size is not determined by the cell switching frequency. Therefore, SiC switching devices are not of special merit here, even when going to higher V_{cell} (i.e., to reduce the number of stacked modules), still Si IGBT would be the selected option [14]. On the other hand, SiC free-wheeling antiparallel diodes are quite interesting in this case since the CHB full bridge is required to handle positive and negative currents (i.e., not only during the dead time). Therefore, SiC diodes can effectively improve the CHB efficiency due to their reduced conduction forward voltage drop as well as lower reverse recovery time compared to Si diodes [27].

DAB: high switching frequencies can provide significant reduction of the size and weight of the converter magnetics and the input and output capacitors. Additionally, going to higher PET module voltage is not achievable using Si devices, unless the switching frequency is reduced to few kHz. Regarding the antiparallel diode, the advantages of using SiC are still a controversial issue [14,28] and, therefore, will be analyzed in this work.

3P4L DC/AC: commutation requirements are not high (in the range of few kHz [29]), and since it interfaces the LVAC grid, high blocking voltages are not required. Therefore, Si devices are a good candidate for this stage.

Considering the above discussion, for this PET, 1.7 kV Si IGBT devices with SiC freewheeling diodes are used for the CHB FB. As for the DAB, 1.2 kV SiC MOSFETs are used and possible enhancements introduced by a SiC antiparallel diode will be discussed in detail in Section 4. Finally, 1.7 kV Si IGBTs are used for the 3P4L converter.

In the next section, the possible achievable enhancements gained by employing SiC in the DAB are discussed highlighting the practical limitations intrinsic to this PET topology.

3.2. Benefits and Practical Limitations of Using SiC MOSFETs for the DAB Converter

The use of SiC MOSFETs for the DAB converter in this PET structure has two main benefits:

1. It enables increasing the switching frequency of the DAB and, therefore, decreasing the size of the HFT [30,31].
2. The high blocking voltages enable increasing V_{cell} and, therefore, for a given V_{acHV}, the number of stacked PET modules can be reduced (see Figure 1) [7,17]. This is also advantageous for the HFT. Since each module handles more power, this leads to a higher transformer power density. In order to see clearly the effect of this, Figure 2 shows the HFT power density as a function of the cell voltage where the PET total power and V_{acHV} are fixed (i.e., the number of PET modules vary). A comparative analysis regarding this relation is previously presented in [17] showing an improvement in the HFT power density as V_{cell} increases.

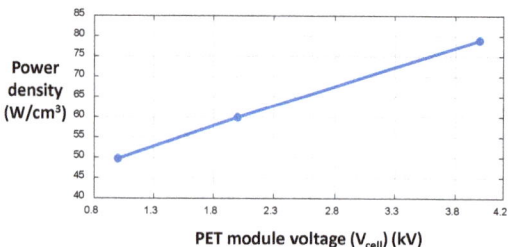

Figure 2. Theoretically calculated High Frequency Transformer (HFT) power density for three designs varying the Power Electronic Transformer (PET) module voltage (V_{cell}) and the handled power [17].

However, these SiC potential benefits may be compromised by certain practical implementation constraints.

On one hand, increasing V_{cell} has several adverse effects. High DC link voltages create practical problems for feeding the control circuitry in each cell. As commercial auxiliary power supplies (APS) do not provide the required isolation [29], each module circuitry has to be supplied from its DC link, where the HV and LV sides have separate APSs [32]. Commercial APSs can be used for voltages under 1 kV, otherwise, custom solutions must be implemented, such as the modular ISOP topology proposed in [29]. Consequently, various aspects must be considered for the selection of V_{cell}.

On the other hand, the size reduction of the HFT, in this particular case, is constrained by the high isolation required by the PET. This isolation imposes minimum clearance distances between windings, which compromises the window utilization factor resulting in a physical limit on further size reduction.

4. SiC-Based DAB Converter

The DAB (see Figure 3) is selected for the intermediate stage of the PET, as it provides galvanic isolation as well as bidirectional power flow [33,34]. The DAB is based on two active bridges interfaced through an HFT, which provides the required galvanic isolation. This converter provides bidirectional operation by controlling the phase shift between the AC voltages generated by both bridges (V_1 and V_2). Also, this converter can have relatively high efficiency due to the soft-switching operation of all the devices at nominal conditions (zero-voltage switching, ZVS) [35,36].

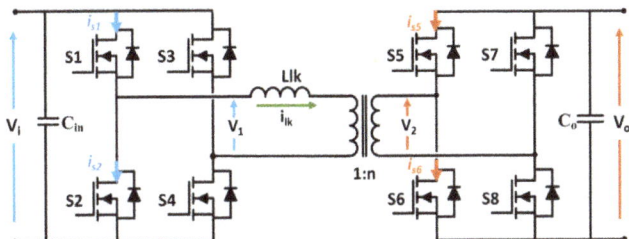

Figure 3. Dual active bridge (DAB) converter schematic.

4.1. SiC Device Selection

Regarding the selection of the SiC devices for the DAB, 6.5, 10, and 15 kV SiC MOSFETs have been reported for laboratory prototypes [37–42], but are far from being a viable commercial alternative yet. Current ratings offered in commercially available 1.7 kV SiC MOSFETs are still limited [23], and their commutation characteristic must be improved. The 1.2 kV SiC devices remain as the most mature technology available in the SiC device market. Consequently, these devices have been selected for the DAB. The DC link voltage is, therefore, set to 800 V.

Seven 1.2 kV SiC commercial MOSFETs were selected; two power modules and five discrete N-channel SiC MOSFETs. A comparative analysis of these devices was carried out in a boost converter operating in continuous conduction mode (CCM) (see Figure 4).

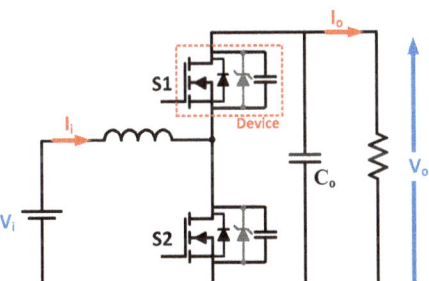

Figure 4. Boost converter test bench schematic.

The boost converter was chosen as a preliminary test bench due to its simplicity and rapid prototyping, but foremost, due to its similar operation to a DAB converter as it has two switching devices in a leg with an inductance connected to the middle point. However, the differences in operation between a CCM boost and a DAB are well understood and, therefore, perspective devices resulting from the first selection stage are then tested in a DAB converter prototype.

The test bench used commercial driver boards from CREE and a commercial FPGA-based controller platform (BASYS2). Tests in the boost converter were done at 2 kW 400/800 V for switching frequencies of 30, 50, and 100 kHz and a dead time of 500 ns. Table 2 summarizes the results. The efficiency is calculated using the input and output DC voltages and currents of the converter measured using digital multimeters. It is observed that all the devices show a high efficiency barely affected by the increase in switching frequency which was increased by a factor of more than three.

Table 2. SiC MOSFET comparative analysis in a boost converter.

Manufacturer	Reference	Package	Rated Current @100 °C (A)	R_{DS} (mΩ)	C_{out} (pF)	Efficiency (%) 30 kHz	Efficiency (%) 50 kHz	Efficiency (%) 100 kHz	Price per Bridge (€)
ST	SCT30N120		34	80	130	97.77	98.16	97.81	100
ROHM	SCH2080KE + SBD	TO-247-3	28	80	175	97.71	98.02	97.69	132
	SCT2080KE		28	80	77	97.82	98.17	97.93	80
	C2M0040120D		40	40	150	97.76	98.10	97.84	128
	C2M0025120D		60	25	220	97.56	97.88	97.42	252
CREE	CAS120M12BM2 + SBD	6-pack 45 mm	59	25	400	97.49	97.54	97.50	412
	CCS050M12CM2 + SBD	Half-bridge 62 mm	138	13	900	96.44	95.8	-	660

Since the performance of all seven MOSFETs is comparable, the selection of the adequate option was mainly based on the size and the price. The specifications for the DAB converter are shown in Table 1. Accordingly, the peak current handled by the devices can be calculated from Equation (1) [35], where T is half the switching period, d is the phase shift, L_{lk} is the leakage inductance, v_o is the output voltage, v_i is the input voltage and n is the HFT turns ratio:

$$I_{p\_lk} = \frac{T}{2L_{lk}}\left(2\frac{v_o}{n}d + v_i - \frac{v_o}{n}\right) \qquad (1)$$

The current peak is calculated for the maximum phase-shift, this is selected to be 0.35 according to [35]. Based on Equation (1), this current is approximately equal to 11 A. The device current rating is selected to be twice the magnitude of the peak current handled by the devices to keep a safety margin. Therefore, the minimum required current rating is 22 A. This eliminates the two modules, and the CREE 60 A discrete, as the size and price are not justified in this case. The remaining four discrete devices are almost equally favored except for the ROHM SCH2080KE, as it includes a SiC Schottky barrier diode (SBD) co-packaged with the MOSFET.

4.2. Antiparallel SBD for a DAB Converter

Observing the efficiency of the boost converter using ROHM SCH2080KE versus for example ROHM SCT2080KE (without an additional SBD), it is consistently higher without SiC SBD. For the boost converter, this is logical as an additional antiparallel diode increases the output capacitance (see Table 2) and since hard-switching occurs, this increases the switching losses. Although, this makes sense and is simple to understand for the boost, for the DAB, it is more complicated as ZVS is implemented. That being the case, it is not valid to make a selection between both devices unless the additional diode behavior is studied for the DAB operation to identify if it improves or worsens its efficiency. This issue is addressed as follows, including: (1) understanding the potential effects introduced by a SiC antiparallel diode in a DAB converter, (2) developing a simple analytical loss model to estimate the possible efficiency improvement introduced by the SiC diode at a certain DAB operating point, and (3) validation of the proposed model using experimental results in a DAB prototype.

The two devices used in the analysis are the ROHM devices (i.e., SCH2080KE and SCT2080KE) as it is the same die but one packed with a SiC antiparallel SBD [22]. Characteristics of both devices are provided in Table 3. The diode forward voltage, $V_{F\text{-diode}}$, is obtained from the datasheet at the value of I_{p_lk} (see Equation (1)) where the employed phase shift is that corresponding to 5 kW.

Table 3. Specifications of the compared devices.

Characteristics		SCH2080KE +SBD	SCT2080KE
$R_{ON\text{-}MOSFET}$		125 mΩ	
E_{off} (R_g = 10 Ω)		110 µJ	120 µJ
E_{on} (R_g = 10 Ω)		330 µJ	275 µJ
Anti-parallel diode	Type	SiC SBD	Body diode
	R_D (Estimated)	45 mΩ	320 mΩ
	V_{knee}	0.85 V	1.4 V
	$V_{F\text{-}Diode}$ (@ I_{p_lk})	1.2 V	4.5 V

4.2.1. Advantages and Disadvantages of an Additional SBD

In order to provide a qualitative preliminary understanding of the possible effects of the SiC SBD on the DAB efficiency, the waveforms of the DAB are shown in Figure 5.

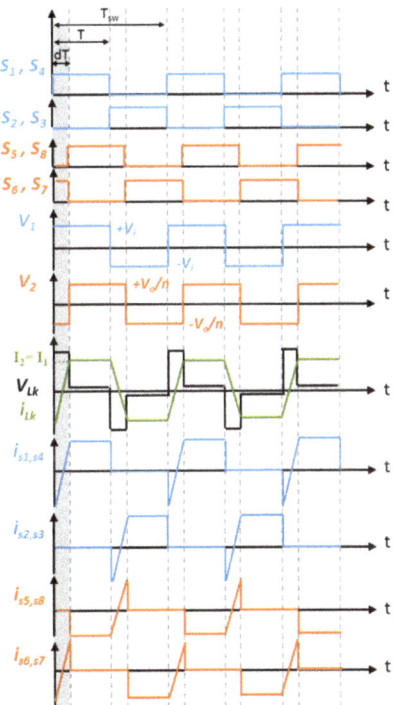

Figure 5. Dual Active Bridge (DAB) converter waveforms.

The potential advantages of using SCH2080KE +SBD in comparison to SCT2080KE are analyzed as follows:

- Since $V_{F\text{-}Diode}$ of the additional SiC SBD is more than three times lower than that of the body diode (see Table 3), the power losses due to diode conduction during the dead times are lower using SCH2080KE +SBD.
- The diode can enter into conduction while the MOSFET is ON. This can be seen from Figure 5, when the current through the MOSFET is negative (i.e., source to drain), a forward voltage drop,

$V_{F\text{-MOSFET}}$, is applied to the diode. If $V_{F\text{-MOSFET}}$ is higher than the diode knee voltage (V_{knee}), then the diode conducts. If this case is true, current is shared between the MOSFET and the diode and, therefore, conduction losses are reduced.

- Unlike boost converter operation, soft switching is adopted in DAB devices at turn ON and, therefore, the additional output capacitance introduced by the additional SiC diode, illustrated in Figure 4, does not significantly penalize the switching losses (see Table 2).

However, these advantages can be compromised by several situations resulting in almost no advantages of having a SiC SBD, these cases are summarized as follows:

- From Figure 6, diodes do not operate during the whole period of the dead time. During (A), the output parasitic capacitances of the MOSFETs (see Figure 4) in one leg are exchanging the voltage (i.e., one is discharging while the other is charging) and during (B), diodes conduct due to circulating currents. In some cases, (B) can tend to zero depending on the value of the current charging/discharging the parasitic capacitors and on the dead time.
- If the current through the MOSFETs is not high enough to produce a $V_{F\text{-MOSFET}} > V_{knee}$, then the diode will never conduct during MOSFET ON time.
- The turn OFF switching losses should be analyzed as the enhancement introduced by the SiC SBD in the conduction losses can, in some cases, be compromised by the increase in turn OFF switching losses (due to the extra capacitance).

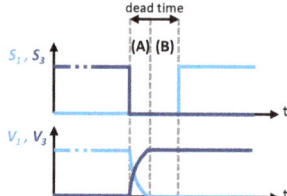

Figure 6. Waveforms of the gates and drain to source voltages of two MOSFETs in one leg during the dead time period.

As a conclusion to the previous discussion, it is important to develop a simple model to determine, for a certain DAB operation point, if an extra antiparallel SiC diode co-packed with the SiC MOSFET is worthy or avoiding it is better. This is introduced in the next section.

4.2.2. Model to Assess the Efficiency Improvement Using a SiC Diode

Diode Conduction Intervals

Regarding diode conduction interval during the dead time, in order to estimate the time duration (B) (see Figure 6), the capacitor charging time during (A) is estimated using Equation (2), where $C_{o(er)}$ is the MOSFET effective output capacitance given by the data sheet and $V_1 = V_i = V_o$. Accordingly, i_{lk} is considered flat ($I_1 = I_2$ in Figure 5) and, therefore, as an approximation, the peak current (I_{p_lk}) is considered to be equal during all the switching transitions.

$$t_{(A)} = 2 \frac{V_1 \, C_{o(er)}}{I_{p\_lk}} \qquad (2)$$

Since $t_{(B)}$ = dead time – $t_{(A)}$, therefore diode operates for 570 ns for the case of the SiC SBD and 588 ns for the body diode. This time represents around 95% of the total dead time (600 ns) and, therefore, the reduction in conduction losses during diode operation is relevant to consider.

Regarding the diode conduction interval during MOSFET ON time, first, the diode characteristics are identified from the MOSFET datasheet (see Table 3). Figure 7 shows the forward voltage drop for both diodes ($V_{F\text{-}Diode}$) as a function of the current through the diode. Additionally, the voltage drop on the MOSFET due to its ON resistance ($V_{F\text{-}MOSFET}$) as a function of the current it is conducting is illustrated.

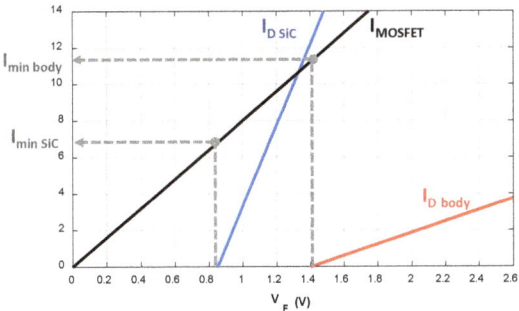

Figure 7. MOSFET voltage drop and diode characteristic curve for the SiC Schottky Barrier Diode (SBD) and the body diode.

It is possible to see that each diode conducts when the current through the MOSFET is above a certain minimum value. It is clearly lower in the case of the SiC SBD. This can be estimated from Equation (3), where $R_{ON\text{-}MOSFET}$ stands for ON-state drain-to-source MOSFET resistance:

$$I_{min} = \frac{V_{knee}}{R_{ON\text{-}MOSFET}} \qquad (3)$$

Based on the data in Table 3, $I_{min\ SiC}$ is 6.8 A while $I_{min\ body}$ is 11.2 A. Since the theoretical calculated $I_{p\_lk} = 9$ A, therefore it is not possible that the body diode enters into conduction while the MOSFET is ON. On the other hand, the SiC diode would conduct when the current exceeds 6.8 A.

Estimation of the Power Losses

Power losses are estimated for both cases of the DAB using MOSFETs with a SiC antiparallel diode (SCH2080KE) and with only the body diode (SCT2080KE). Theoretical power loss estimation will be compared to the experimental efficiency results presented in the next section.

First, the periods where the diode can conduct (when it is forward biased) are identified as shown in Figure 8. It can be seen that, the diodes of all the ON MOSFETs can conduct during (1) and (4) as the current through all the devices is negative (see Figure 5), while during (3) and (6) only the diodes of the secondary bridge can conduct.

Accordingly, to simplify the power loss estimation the following assumptions are made:

- As $I_{min\ SiC}$ is around 75% of $I_{pk\text{-}lk}$, therefore, it is assumed that the diode operates only during period (3) and (6) when the peak current is passing through the devices (periods (1) and (4) are neglected).
- The power loss estimation is performed only during the periods with different operation for the case of the SiC SBD and the body diode. In other words, only the difference in losses between both cases is considered. These intervals are: (3), (6) and the dead times.

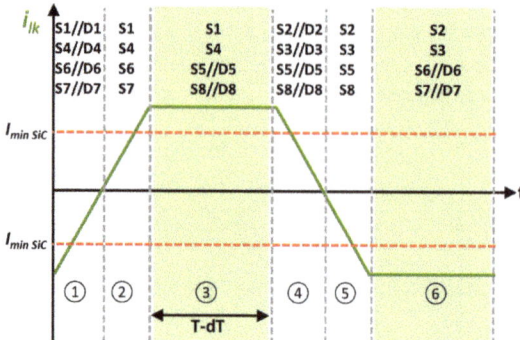

Figure 8. Current through the power transfer inductance of the Dual Active Bridge (DAB) showing periods where the antiparallel diode can enter into conduction mode during MOSFETs conduction.

During intervals (3) and (6), the conduction and switching losses of all the devices are estimated, while during the dead times, only the diodes conduction losses are estimated.

(a) *Losses during dead time:*

During one switching period (T_{sw}), four intervals of dead time take place. During each interval two diodes conduct. Accordingly, the diodes total conduction losses are estimated using Equation (4), where $V_{F\text{-Diode}}$ is the diode forward voltage at I_{p_lk} obtained from the diode characteristic curve given by the datasheet (linearized in Figure 7).

$$P_{deadtime} = 8 \cdot V_{F-Diode} \cdot I_{p\_lk} \cdot \frac{t_{(B)}}{T_{sw}} \qquad (4)$$

(b) *Losses during interval (3) and (6) (Figure 8):*

Losses during these two intervals are divided into: (1) turn OFF switching losses (P_{sw}) and (2) MOSFET and diode conduction losses (P_{MOSFET} and P_{Diode}). The switching losses are straightforward, estimated from MOSFET E_{OFF} using Equation (5).

$$P_{sw} = 8 \cdot E_{OFF} \cdot f_{sw} \qquad (5)$$

On the other hand, conduction losses are not straight forward. As noticed from Figure 8, in both (3) and (6) intervals, the primary bridge devices are different from the secondary ones. For example, during (3), for the primary, only the MOSFETs conduct (i.e., all the current flows through the MOSFETs, S1 & S4). In this case, four MOSFETs of the primary bridge conduct during (3) and (6) (two in each interval). Therefore, the total primary conduction losses can be easily estimated using Equation (6).

$$P_{MOSFET-prim} = 4 \cdot I_{p\_lk}^2 \cdot R_{ON-MOSFET} \cdot \frac{T - dT}{T_{sw}} \qquad (6)$$

However, for the secondary, the current is shared between the diode and the MOSFET (S5, D5 and S8, D8). Accordingly, the current conducted by each element has to be estimated. Since the voltage drop on the MOSFET is equal to the diode $V_{F\text{-Diode}}$, then:

$$I_M \cdot R_{ON-MOSFET} = V_{knee} + I_D \cdot R_D \qquad (7)$$

where I_M and I_D are the currents through the MOSFET and the diode respectively and R_D is the diode dynamic resistance.

Since the current is shared by both elements, then, I_M and I_D can be obtained from Equations (7) and (8):

$$I_M + I_D = I_{p\_lk} \tag{8}$$

Finally, the total conduction losses of the secondary bridge during T_{sw} can be calculated from Equations (9) and (10):

$$P_{MOSFET-sec} = 4 \cdot I_M{}^2 \cdot R_{ON-MOSFET} \cdot \frac{T - dT}{T_{sw}} \tag{9}$$

$$P_{Diode-sec} = 4 \cdot I_D \cdot V_{F-Diode}|_{I_D} \cdot \frac{T - dT}{T_{sw}} \tag{10}$$

The values of the power loss components estimated in this section are summarized in Table 4 where the enhancement to the DAB efficiency is estimated to be approximately 0.22% given the rated power is 5 kW.

Table 4. Theoretically-estimated power losses.

Interval	Losses		SCH2080KE +SBD	SCT2080KE
Dead time (600 ns)	Diode conduction ($P_{deadtime}$)		1.6 W	6.2 W
(3) & (6)	Switching (P_{sw})		26.4 W	28.8 W
	Conduction	$P_{MOSFET-prim}$	17.4 W	17.4 W
		$P_{MOSFET-sec}$	10.4 W	17.4 W
		$P_{Diode-sec}$	2.8 W	-
	Total losses		58.6 W	69.8 W

4.2.3. Experimental Validation of the Proposed Loss Model

A DAB test bench was constructed and tests were performed at the nominal operation defined in Table 1 and the results are summarized in Table 5.

Table 5. Experimental results of the Dual Active Bridge (DAB) with and without SiC Schottky Barrier Diode (SBD).

Characteristics	SCH2080KE +SBD	SCT2080KE
Antiparallel diode	SiC SBD	Body diode
Efficiency (η)	98.1%	97.8%
Phase-shift (d)	0.29	
Dead time	600 ns	
I_{p_lk}	Calculated from Equation (1): 9 A Measured (experimental): 9.8 A	

It was observed that the measured efficiency of the DAB converter composed of MOSFETs with an additional SiC antiparallel SBD is higher than that with only the body diode. The enhancement observed from the experimental results is around 0.3%, which validates the calculation introduced previously resulting in 0.22%. These results validate the proposed hypothesis and approach.

5. Experimental Results of the PET Module

The developed full-scale PET module is shown in Figure 9. Its structure is that schematically shown in Figure 1b. Table 6 summarize the main components of the PET module.

Figure 9. The developed Power Electronic Transformer (PET) module.

Table 6. Main Power Electronic Transformer (PET) module components.

Element	Characteristics	Reference
DAB	ROHM 1.2 kV, 28 A SiC MOSFET CREE 2-channel drivers	SCH2080KE CGD15HB62P1
CHB	Infineon 1.7 kV Si IGBT + SiC diode Infineon two-channel drivers	Not commercial 2ED300C17-S
Controller	Xilinx FPGA (Spartan 3E) custom board	XC3S250E-4TQG144I

5.1. Developed HFT

One of the key aspects in the design of this HFT is the high galvanic isolation required between its primary and secondary sides, being 24 kV in this PET. This presents significant challenges compared to the ones considered in literature [43,44]. Moreover, the DAB power transfer inductance (L_{lk}) is magnetically integrated in the HFT by making use of its series leakage inductance [43]. This reduces size and cost but, on the other hand, imposes additional constrains on the HFT design due to the required accuracy of L_{lk}.

The HFT design is, therefore, a tradeoff between four variables: transferred power capability, temperature rise (i.e., losses), size and cost. Several design iterations are performed to achieve the required isolation with an acceptable tradeoff between these variables. The HFT design was previously presented in [45]. The main experimental validation tests are provided in this work for completeness.

Figure 10a shows laboratory developed HFT for test purposes and Figure 10b shows the final factory encapsulated HFT. A UU core structure is used typically selected in literature for separate winding [46,47]. The core ferrite material is selected to be Ferroxcube® 3C90 based on the DAB switching frequency [43]. An epoxy resin providing 15 kV/mm and exhibiting acceptable thermal conductance of 3 W/mK was used for encapsulation.

Figure 11 shows a schematic representation of the HFT design from ANSYS PEmag® software (Canonsburg, PA, USA).

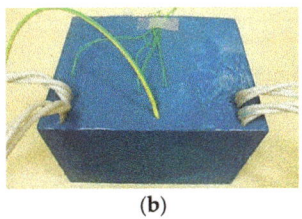

(a)   (b)

Figure 10. High Frequency Transformer (HFT) prototype. (**a**) Non-encapsulated laboratory developed using 3D printed bobbins. (**b**) Encapsulated final HFT.

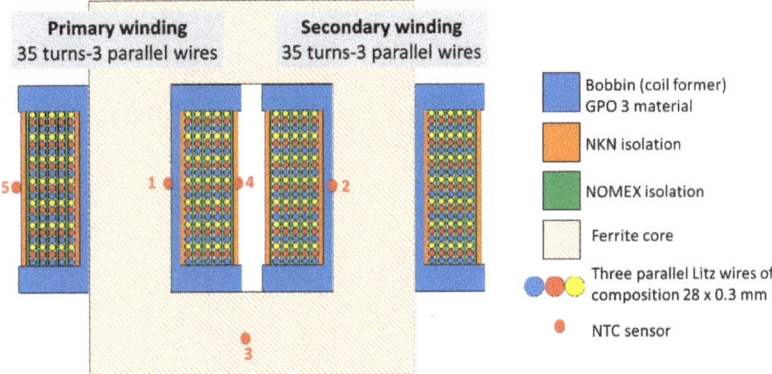

Figure 11. High Frequency Transformer (HFT) design taken from ANSYS PEmag®.

To verify the achieved isolation, a high potential test is done for both porotypes and results are compared. A Hipot tester is used to apply a voltage potential of up to 24 kV between both windings. The leakage current flowing from the winding with the higher potential, through air/resin, is recorded and shown in Figure 12. A significant diversion between both prototypes is clear at higher voltage potentials. Having lower leakage currents means that successive partial discharge due to high dv/dt is avoided which would lead to eventual insulation breakdown [48].

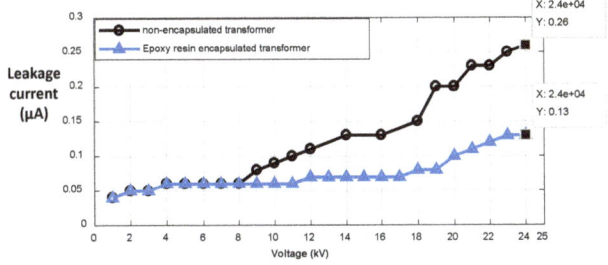

Figure 12. Experimental results. Leakage current measured between High Frequency Transformer (HFT) windings, as a function of the voltage difference between them.

To locate the hottest spot and verify its temperature rise, five NTC sensors were mounted inside the HFT (see Figure 11). The location of this spot differs in encapsulated (NTC 1) and non-encapsulated prototypes (NTC 4). The temperature profile of the encapsulated HFT hottest spot during a four-hour

DAB test at rated operation is shown in Figure 13. The steady state temperature, under natural convection, was recorded to be 60 °C.

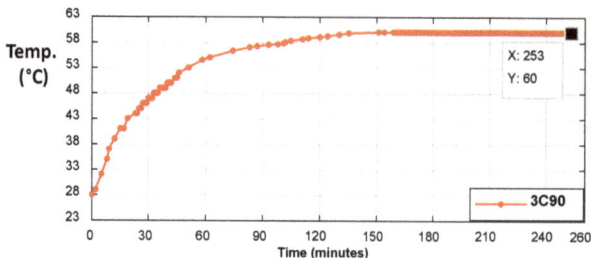

Figure 13. Experimental results. Hottest spot temperature profile for the encapsulated High Frequency Transformer (HFT) using 3C90 ferrite core measured using a Negative Temperature Coefficient (NTC) sensor.

A summary of the final HFT design is shown in Table 7.

Table 7. High Frequency Transformer (HFT) design summary.

Quantity	Value
Isolation	24 kV
Core structure	UU100/57/25
Ferrite material	Ferroxcube® 3C90
Encapsulation	Epoxy resin (15 kV/mm) (ROYAPOX 912 THC/2)
Bobbins	GPO-3 (20 kV/mm)
Number of turns (N)	35
Temperature rise at NTC 4	40 °C

5.2. Experimental Validation

Figure 14 shows the performed test connection diagram. A DC power supply is connected to the LVDC side to provide V_{dcLV} and the power is transferred from FB2 to FB1. The DAB current control regulates the transferred power using single phase shift (SPS) modulation while the CHB full bridge regulates the cell capacitor voltage (V_{cell}).

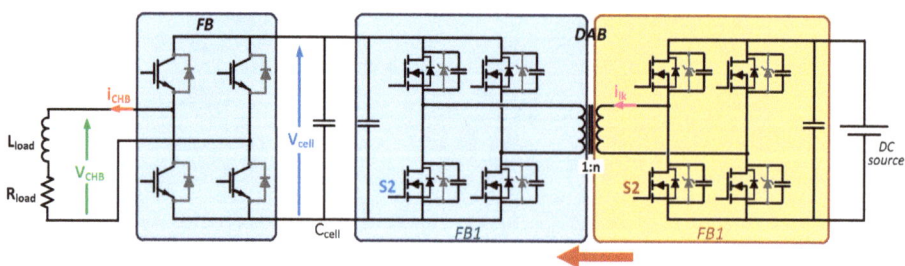

Figure 14. Schematic of the Power Electronic Transformer (PET) module during the performed experimental test.

Figure 15 shows the experimental results for the PET module nominal operation (described in Table 1). Figure 15a shows the DAB waveforms, where FB2 gates are leading FB1 gates and the phase shift between both controls the magnitude of the transferred power [35].

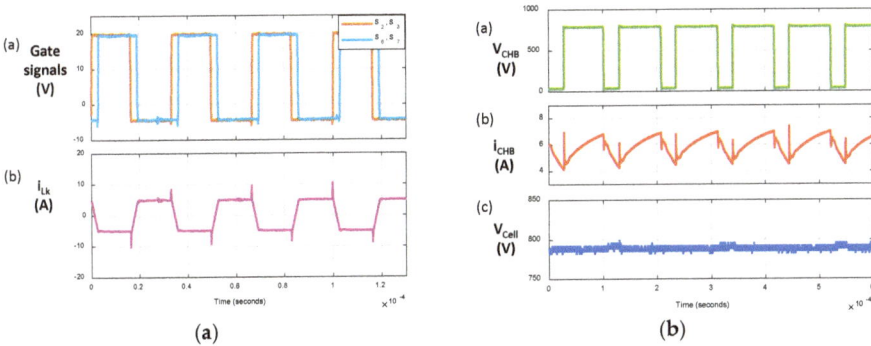

Figure 15. Experimental results. (a) Dual Active Bridge (DAB) waveforms. Two gate signals of DAB FB1 and FB2 and leakage inductance current (i_{lk}). (b) Cascaded H-Bridge (CHB) full bridge waveforms. CHB full bridge output voltage (V_{CHB}), CHB full bridge output current (i_{CHB}), and the cell capacitor voltage (V_{cell}).

Figure 15b shows the CHB waveforms, where V_{CHB} and i_{CHB} are the CHB full-bridge output voltage and current respectively. V_{cell} is regulated at 800 V by the CHB voltage control.

6. Conclusions

This paper analyzes the use of SiC devices in three-stage modular PETs. It has been shown that SiC MOSFETs can be especially advantageous in the isolation stage, as they combine high blocking voltage and high switching frequencies, leading to higher efficiency, size reduction of the isolation transformer, and higher power density. On the other hand, limited merit is achieved using SiC in the front end AC/DC and the LV side DC/AC converters. However, this partially depends on the application of the PET and the selected topology.

The paper details the selection of SiC MOSFETs for the DAB DC/DC isolation stage experimentally comparing different commercial devices. The benefits of a SiC antiparallel SBD is investigated. It is concluded that the SiC diode improves the efficiency of the DAB as it reduces the total conduction losses of the device. However, this improvement strongly depends on two critical design aspects: the employed dead time interval and the time interval during which the current is driven by both the MOSFET and the diode at the same time (due to the voltage drop in the MOSFET on resistance directly biasing the diode). Generally, the efficiency improvement can be more relevant for a flatter inductor current shape and when the DAB is working around its nominal operation point. A loss model was presented to assess the introduced efficiency improvement and was experimentally validated.

Experimental results showing the operation of the full-scale DAB converter at rated conditions as part of the PET module are provided.

Author Contributions: Conceptualization, A.R., M.A. and F.B.; methodology, A.R., M.A. and F.B.; software, M.S.; validation, M.S. and M.R.R.; formal analysis, M.S. and M.R.R.; investigation, M.S. and M.R.R.; writing—review and editing, M.S., M.R.R., A.R., M.A. and F.B.; supervision, A.R., M.A. and F.B.; funding acquisition, A.R., M.A. and F.B. All authors have read and agreed to the published version of the manuscript.

Funding: This work was supported by the Spanish Government under projects MCIU-19-RTI2018-099682-A-I00, FC-GRUPIN-IDI/2018/000179, and PAPI-18-PF-10 and by the European Commission FP7 Large Project NMP3-LA-2013-604057, under grant UE-14-SPEED-604057.

Conflicts of Interest: The authors declare no conflict of interest.

Nomenclature

V_{acHV}	HVAC grid voltage
V_{acLV}	LVAC grid voltage
V_{dcLV}	LVDC link voltage
C_{dcLV}	LVDC link capacitor
V_{cell}	Cell capacitor voltage
C_{cell}	Cell capacitor
f_{sw}	Switching frequency
T	Half the switching period of the DAB
d	Phase shift of the DAB represented as a ratio of T
L_{lk}	DAB leakage inductance
V_{knee}	Diode knee voltage
$V_{F\text{-}diode}$	Diode forward voltage
$I_{pk\text{-}lk}$	Peak current through the leakage inductance
$R_{ON\text{-}MOSFET}$	MOSFET ON resistance
$C_{o(er)}$	MOSFET effective output capacitance

References

1. van der Merwe, J.W.; Mouton, H.d.T. The solid-state transformer concept: A new era in power distribution. In Proceedings of the AFRICON 2009, Nairobi, Kenya, 23–25 September 2009; pp. 1–6.
2. McMurray, W. Power Converter Circuits Having a High-Frequency Link. U.S. Patent 3517300, 23 June 1970.
3. Ronan, E.R.; Sudhoff, S.D.; Glover, S.F.; Galloway, D.L. A power electronic-based distribution transformer. IEEE Trans. Power Del. 2002, 17, 537–543. [CrossRef]
4. Rothmund, D.; Ortiz, G.; Guillod, T.; Kolar, J.W. 10kV SiC-based isolated DC-DC converter for medium voltage-connected Solid-State Transformers. In Proceedings of the IEEE Applied Power Electronics Conference and Exposition (APEC), Charlotte, NC, USA, 15–19 March 2015; pp. 1096–1103.
5. Millán, J.; Godignon, P.; Perpiñà, X.; Pérez-Tomás, A.; Rebollo, J. A Survey of Wide Bandgap Power Semiconductor Devices. IEEE Trans. Power Electron. 2014, 29, 2155–2163. [CrossRef]
6. Millán, J.; Friedrichs, P.; Mihaila, A.; Soler, V.; Rebollo, J.; Banu, V.; Godignon, P. High-voltage SiC devices: Diodes and MOSFETs. In Proceedings of the International Semiconductor Conference (CAS), Sinaia, Romania, 12–14 October 2015; pp. 11–18.
7. Dong, D.; Agamy, M.; Bebic, J.Z.; Chen, Q.; Mandrusiak, G. A Modular SiC High-Frequency Solid-State Transformer for Medium-Voltage Applications: Design, Implementation, and Testing. IEEE J. Emerg. Sel. Top. Power Electron. 2019, 7, 768–778. [CrossRef]
8. Raju, R.; Dame, M.; Steigerwald, R. Solid-state transformer using silicon carbide-based modular building blocks. In Proceedings of the IEEE International Conference of Power Electronics and Drive Systems (PEDS), Honolulu, HI, USA, 12–15 December 2017; pp. 1–7.
9. Yao, R.; Zheng, Z. Design and control of a SiC-based three-stage AC to DC power electronic transformer for electric traction applications. J. Eng. 2019, 2019, 2947–2952. [CrossRef]
10. Wang, G.; Huang, X.; Wang, J.; Zhao, T.; Bhattacharya, S.; Huang, A.Q. Comparisons of 6.5kV 25A Si IGBT and 10-kV SiC MOSFET in Solid-State Transformer application. In Proceedings of the IEEE Energy Conversion Congress and Exposition (ECCE), Atlanta, GA, USA, 12–16 September 2010; pp. 100–104.
11. Madhusoodhanan, S.; Tripathi, A.; Patel, D.; Mainali, K.; Kadavelugu, A.; Hazra, S.; Bhattacharya, S.; Hatua, K. Solid-State Transformer and MV Grid Tie Applications Enabled by 15 kV SiC IGBTs and 10 kV SiC MOSFETs Based Multilevel Converters. IEEE Trans. Ind. Appl. 2015, 51, 3343–3360. [CrossRef]
12. Hatua, K.; Dutta, S.; Tripathi, A.; Baek, S.; Karimi, G.; Bhattacharya, S. Transformerless intelligent power substation design with 15 kV SiC IGBT for grid interconnection. In Proceedings of the IEEE Energy Conversion Congress and Exposition (ECCE), Phoenix, AZ, USA, 17–22 September 2011; pp. 4225–4232.
13. Abu-Siada, A.; Budiri, J.; Abdou, A.F. Solid State Transformers Topologies, Controllers, and Applications: State-of-the-Art Literature Review. Electronics 2018, 7, 298. [CrossRef]

14. Evans, N.M.; Lagier, T.; Pereira, A. A preliminary loss comparison of solid-state transformers in a rail application employing silicon carbide (SiC) MOSFET switches. In Proceedings of the 8th IET International Conference on Power Electronics, Machines and Drives (PEMD), Glasgow, UK, 19 April 2016; pp. 1–6.
15. Lopez, M.; Briz, F.; Zapico, A.; Rodriguez, A.; Diaz-Reigosa, D. Control strategies for MMC using cells with power transfer capability. In Proceedings of the IEEE Energy Conversion Congress and Exposition (ECCE), Montreal, QC, Canada, 20–24 September 2015; pp. 3570–3577.
16. Cuartas, J.M.; de la Cruz, A.; Briz, F.; Lopez, M. Start-up, functionalities and protection issues for CHB-based solid state transformers. In Proceedings of the IEEE International Conference on Environment and Electrical Engineering and IEEE Industrial and Commercial Power Systems Europe (EEEIC/I&CPS Europe), Milan, Italy, 6–9 June 2017; pp. 1–5.
17. López, M.; Briz, F.; Saeed, M.; Arias, M.; Rodríguez, A. Comparative analysis of modular multiport power electronic transformer topologies. In Proceedings of the IEEE Energy Conversion Congress and Exposition (ECCE), Milwaukee, WI, USA, 18–22 September 2016; pp. 1–8.
18. Briz, F.; Lopez, M.; Rodriguez, A.; Arias, M. Modular Power Electronic Transformers: Modular Multilevel Converter versus Cascaded H-Bridge Solutions. *IEEE Ind. Electron. Mag.* **2016**, *10*, 6–19. [CrossRef]
19. Briz, F.; López, M.; Zapico, A.; Rodríguez, A.; Díaz-Reigosa, D. Operation and control of MMCs using cells with power transfer capability. In Proceedings of the IEEE Applied Power Electronics Conference and Exposition (APEC), Charlotte, NC, USA, 15–19 March 2015; pp. 980–987.
20. European Union. "Silicon Carbide Power Technology for Energy Efficient Devices (SPEED)", Ref. FP7-NMP3-LA-2013-604057, EU–FP7, Large Scale Integrating Collaborative Research Project. Available online: https://cordis.europa.eu/project/id/604057 (accessed on 4 March 2020).
21. Wensong, Y. *Wide Band-Gap Devices for Solid State Transformer Applications*; Tutorial; FREEDM Systems Center: Raleigh, NC, USA, 2017.
22. ROHM Semiconductor Electronics Industry Company. Available online: www.rohm.com (accessed on 2 March 2020).
23. Wolfspeed, Inc. Available online: www.wolfspeed.com (accessed on 2 March 2020).
24. Zhang, L.; Yuan, X.; Wu, X.; Shi, C.; Zhang, J.; Zhang, Y. Performance Evaluation of High-Power SiC MOSFET Modules in Comparison to Si IGBT Modules. *IEEE Trans. Power Electron.* **2019**, *34*, 1181–1196. [CrossRef]
25. Malinowski, M.; Gopakumar, K.; Rodriguez, J.; Perez, M.A. A Survey on Cascaded Multilevel Inverters. *IEEE Trans. Ind. Electron.* **2010**, *57*, 2197–2206. [CrossRef]
26. Rodriguez, J.; Lai, J.-S.; Peng, F.Z. Multilevel inverters: A survey of topologies, controls, and applications. *IEEE Trans. Ind. Electron.* **2002**, *49*, 724–738. [CrossRef]
27. Sahib, A.; Abdulbaqi, I.M.; Ahmed, A.H. Analysis of a Four-Leg Inverter Feeding a Nonlinear Load using SVPWM Technique. In Proceedings of the Second Al-Sadiq International Conference on Multidisciplinary in IT and Communication Science and Applications (AIC-MITCSA), Baghdad, Iraq, 30–31 December 2017; pp. 19–24.
28. Martin, D.; Killeen, P.; Curbow, W.A.; Sparkman, B.; Kegley, L.E.; McNutt, T. Comparing the switching performance of SiC MOSFET intrinsic body diode to additional SiC schottky diodes in SiC power modules. In Proceedings of the IEEE 4th Workshop on Wide Bandgap Power Devices and Applications (WiPDA), Fayetteville, AR, USA, 7–9 November 2016; pp. 242–246.
29. Rodriguez, A.; Rogina, M.R.; Saeed, M.; Lamar, D.G.; Arias, M.; Lopez, M.; Briz, F. Auxiliary power supply based on a modular ISOP flyback configuration with very high input voltage. In Proceedings of the IEEE Energy Conversion Congress and Exposition (ECCE), Milwaukee, WI, USA, 18–22 September 2016; pp. 1–7.
30. Huber, J.E.; Kolar, J.W. Solid-State Transformers: On the Origins and Evolution of Key Concepts. *IEEE Ind. Electron. Mag.* **2016**, *10*, 19–28. [CrossRef]
31. McLyman, C.W.T. *High Reliability Magnetic Devices: Design and Fabrication*; Marcel Dekker: New York, NY, USA, 2002.
32. Saeed, M.; Cuartas, J.M.; Rodriguez, A.; Arias, M.; Briz, F. Energization and Start-Up of CHB-Based Modular Three-Stage Solid-State Transformers. *IEEE Trans. Ind. Appl.* **2018**, *54*, 5483–5492. [CrossRef]
33. de Doncker, R.W.; Divan, R.W.; Kheraluwala, M.H. A three-phase soft-switched high power-density dc/dc converter for high-power applications. *IEEE Trans. Ind. Appl.* **1991**, *27*, 63–73. [CrossRef]
34. Kheraluwala, M.H.; Gascoigne, R.W.; Divan, D.M.; Baumann, E.D. Performance characterization of a high-power dual active bridge dc-to-dc converter. *IEEE Trans. Ind. Appl.* **1992**, *28*, 1294–1301. [CrossRef]

35. Rodríguez, A.; Sebastian, J.; Lamar, D.G.; Hernando, M.M.; Vazquez, A. An overall study of a dual active bridge for bidirectional DC/DC conversion. In Proceedings of the IEEE Energy Conversion Congress & Exposition (ECCE), Atlanta, GA, USA, 12–16 September 2010; pp. 1129–1135.
36. Calderon, C.; Barrado, A.; Rodriguez, A.; Alou, P.; Lazaro, A.; Fernandez, C.; Zumel, P. General Analysis of Switching Modes in a Dual Active Bridge with Triple Phase Shift Modulation. *Energies* 2018, *11*, 2419. [CrossRef]
37. Wang, G.; Huang, A.Q.; Wang, F.; Song, X.; Ni, X.; Ryu, S.H.; Grider, D.; Schupbach, M.; Palmour, J. Static and dynamic performance characterization and comparison of 15 kV SiC MOSFET and 15 kV SiC n-IGBTs. In Proceedings of the IEEE 27th International Symposium on Power Semiconductor Devices & IC's (ISPSD), Hong Kong, China, 10–14 May 2015; pp. 229–232.
38. Kumar, A.; Parashar, S.; Sabri, S.; van Brunt, E.; Bhattacharya, S.; Veliadis, V. Ruggedness of 6.5 kV, 30 a SiC MOSFETs in extreme transient conditions. In Proceedings of the IEEE 30th International Symposium on Power Semiconductor Devices and ICs (ISPSD), Chicago, IL, USA, 13–17 May 2018; pp. 423–426.
39. Wang, L.; Zhu, Q.; Yu, W.; Huang, A.Q. A medium voltage bidirectional DC-DC converter combining resonant and dual active bridge converters. In Proceedings of the IEEE Applied Power Electronics Conference and Exposition (APEC), Charlotte, NC, USA, 15–19 March 2015; pp. 1104–1111.
40. Wang, F.; Wang, G.; Huang, A.; Yu, W.; Ni, X. Design and operation of A 3.6kV high performance solid state transformer based on 13kV SiC MOSFET and JBS diode. In Proceedings of the IEEE Energy Conversion Congress and Exposition (ECCE), Pittsburgh, PA, USA, 14–18 September 2014; pp. 4553–4560.
41. Pala, V.; Brunt, E.V.; Cheng, L.; O'Loughlin, M.; Richmond, J.; Burk, A.; Allen, S.T.; Grider, D.; Palmour, J.W.; Scozzie, C.J. 10 kV and 15 kV silicon carbide power MOSFETs for next-generation energy conversion and transmission systems. In Proceedings of the IEEE Energy Conversion Congress and Exposition (ECCE), Pittsburgh, PA, USA, 14–18 September 2014; pp. 449–454.
42. Yun, C.-G.; Cho, Y. Active Hybrid Solid State Transformer Based on Multi-Level Converter Using SiC MOSFET. *Energies* 2019, *12*, 66. [CrossRef]
43. Mu, M.; Xue, L.; Boroyevich, D.; Hughes, B.; Mattavelli, P. Design of integrated transformer and inductor for high frequency dual active bridge GaN Charger for PHEV. In Proceedings of the IEEE Applied Power Electronics Conference and Exposition (APEC), Charlotte, NC, USA, 15–19 March 2015; pp. 579–585.
44. Hoang, K.D.; Wang, J. Design optimization of high frequency transformer for dual active bridge DC-DC converter. In Proceedings of the XXth International Conference on Electrical Machines (ICEM), Marseille, France, 2–5 September 2012; pp. 2311–2317.
45. Saeed, M.; Rogina, M.R.; Lopez, M.; Rodriguez, A.; Arias, M.; Briz, F. Design and construction of a DAB using SiC MOSFETs with an isolation of 24 kV for PET applications. In Proceedings of the IEEE 19th European Conference on Power Electronics and Applications (EPE'17 ECCE Europe), Warsaw, Poland, 11–14 September 2017; pp. P.1–P.10.
46. Ortiz, G.; Biela, J.; Kolar, J.W. Optimized design of medium frequency transformers with high isolation requirements. In Proceedings of the IEEE IECON—36th Annual Conference on Industrial Electronics Society, Glendale, AZ, USA, 7–10 November 2010; pp. 631–638.
47. Zhao, S.; Li, Q.; Lee, F.C. High frequency transformer design for modular power conversion from medium voltage AC to 400V DC. In Proceedings of the IEEE Applied Power Electronics Conference and Exposition (APEC), Tampa, FL, USA, 26–30 March 2017; pp. 2894–2901.
48. Chen, Q.; Raju, R.; Dong, D.; Agamy, M. High Frequency Transformer Insulation in Medium Voltage SiC enabled Air-cooled Solid-State Transformers. In Proceedings of the IEEE Energy Conversion Congress and Exposition (ECCE), Portland, OR, USA, 23–27 September 2018; pp. 2436–2443.

© 2020 by the authors. Licensee MDPI, Basel, Switzerland. This article is an open access article distributed under the terms and conditions of the Creative Commons Attribution (CC BY) license (http://creativecommons.org/licenses/by/4.0/).

Article

Analysis of Intrinsic Switching Losses in Superjunction MOSFETs Under Zero Voltage Switching

Maria R. Rogina [1,*], Alberto Rodriguez [1], Diego G. Lamar [1], Jaume Roig [2], German Gomez [2] and Piet Vanmeerbeek [2]

1. Electrical, Electronic, Computers and Systems Engineering Department, University of Oviedo, 33204 Gijón, Spain; rodriguezalberto@uniovi.es (A.R.); gonzalezdiego@uniovi.es (D.G.L.)
2. On Semiconductor, 9700 Oudenaarde, Belgium; Jaume.Roig@onsemi.com (J.R.); german.gomez@onsemi.com (G.G.); Piet.Vanmeerbeek@onsemi.com (P.V.)
* Correspondence: rodriguezrmaria@uniovi.es;

Received: 7 February 2020; Accepted: 26 February 2020; Published: 2 March 2020

Abstract: Switching losses of power transistors usually are the most relevant energy losses in high-frequency power converters. Soft-switching techniques allow a reduction of these losses, but even under soft-switching conditions, these losses can be significant, especially at light load and very high switching frequency. In this paper, hysteresis and energy losses are shown during the charge and discharge of the output capacitance (C_{OSS}) of commercial high voltage Superjunction MOSFETs. Moreover, a simple methodology to include information about these two phenomena in datasheets using a commercial system is suggested to manufacturers. Simulation models including C_{OSS} hysteresis and a figure of merit considering these intrinsic energy losses are also proposed. Simulation and experimental measurements using an LLC resonant converter have been performed to validate the proposed mechanism and the usefulness of the proposed simulation models.

Keywords: soft-switching; Superjunction MOSFET; LLC resonant converter; zero voltage switching; C_{OSS} hysteresis; C_{OSS} intrinsic energy losses

1. Introduction

During the last 15 years, the acceptance of resonant converters in the industry application market has been massive, especially regarding adapters, flat panel TVs, electric and hybrid vehicle (EV/HEV), datacenters, and photovoltaic (PV) inverters, among others [1–3] (Figure 1). Besides, new markets and research centers are focusing on moving to higher frequencies to obtain further advantages and gaining power density, taking the present technologies in semiconductors to their physical limit. This is the case of gallium nitride (GaN) and silicon carbide (SiC) technologies, which are thought to be used in the market of resonant converters for low power and high-frequency applications, besides other well-known high-power applications.

A resonant topology operating at a high switching frequency and zero voltage switching (ZVS) provides high power density and is commonly chosen for the previously mentioned applications. The primary side power transistors used in a resonant converter must comply with high-voltage and high-frequency requirements, and need to be properly selected to provide good performance. However, the information given by the manufacturers of these transistors is not usually enough to calculate all the existing energy losses.

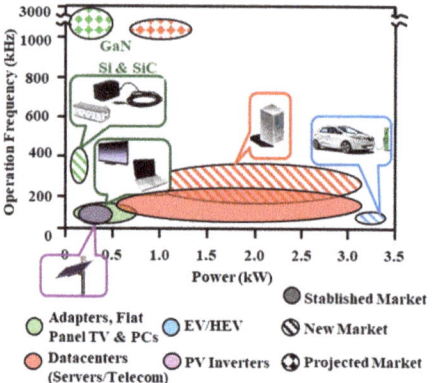

Figure 1. Markets in which resonant converters are used, for different ranges of frequencies (kHz) and power (kW). EV/HEV, electric and hybrid vehicle; PV, photovoltaic.

The parasitic output capacitance (C_{OSS}) of the power transistors has an important role in energy losses, even under ZVS conditions. Traditional switching losses models are not valid for very high switching frequencies. In the work of [4], significant energy dissipation in the process of charging and discharging C_{OSS} of Superjunction MOSFET (SJ-MOSFET) while the gate is shorted to the source is observed. In another paper [5], these intrinsic energy losses (E_i) are measured and compared in different power switches, including Silicon SJ-MOSFETs, GaN cascode, SiC cascode, and SiC MOSFETs. These E_i cannot be eliminated by using ZVS and set an upper limit for the switching frequency of the converters. Similar losses are presented in the work of [6], where energy dissipation during the charging and discharging of the junction capacitance of SiC Schottky diodes is evaluated. In the work of [7], the E_i are included for the determination of soft-switching losses of 10 kV SiC MOSFET modules. Calorimetric measurements are used to evaluate these losses (based on the charge and discharge of the C_{OSS}, and especially of the antiparallel junction barrier Schottky diode). In the work of [8], the variation of E_i with dV/dt is evaluated at very high switching frequency (1–35 MHz) in silicon (Si) and wide-bandgap active devices.

In the work of [9], these E_i are related to a significant hysteresis exhibited by the C_{OSS} of some of the most advanced SJ-MOSFETs. In a further paper [10], the physical mechanism responsible for this C_{OSS} hysteresis is briefly shown by means of mixed-mode simulations. Finally, mixed-mode simulations are also proposed to analyze E_i and the cause of the C_{OSS} hysteresis in different SJ-MOSFETs in the work of [11].

The authors of this paper have previously analyzed the C_{OSS} hysteresis and its related switching losses (including E_i) for different dead-times of three generations of SJ-MOSFETs in an LLC resonant converter in the work of [12]. Moreover, they provide a guideline to select SJ-MOSFETs of different manufacturers to improve the efficiency of this converter in a wide power range in the work of [13].

In this paper, a simple methodology is suggested to manufacturers to include information related to the C_{OSS} hysteresis and E_i of power devices in their datasheets. These data will be useful to select the optimum devices in high-frequency and soft-switching applications. Moreover, a spice model including the C_{OSS} hysteresis and a figure of merit (FoM) including E_i are proposed in Section 2 and 3, respectively. Both proposals are validated using simulation and experimental results of an LLC resonant converter in Section 4.

2. Spice Model Including C_{OSS} Hysteresis Effect

In order to design a resonant converter with low cost and high efficiency and power density, special attention is crucial during the selection of the high-voltage (~600 V) silicon SJ-MOSFET device needed.

However, even if the high-voltage SJ-MOSFETs are selected based on major vendors recommendations for soft-switching applications in resonant converters, they present different values of E_i. E_i might not seem so significant for hard-switching conditions, but it can make the difference under soft-switching operation, especially for low and medium load demands, where conduction power losses are lower and switching losses are relevant because of the high-frequency operation.

E_i is intrinsic to the structure of SJ-MOSFETs, as it is briefly reproduced in Figure 2a–c and explained in detail in the work of [12], where a physical relationship between C_{OSS} hysteresis and E_i was demonstrated, elucidating the existence of energy losses during the charge and discharge of C_{OSS}. Holes and electrons (h+ and e-, respectively, in Figure 2), flowing in parallel to the capacitance, originate stucked charges between the pillars that need to be removed through highly resistive paths that may vary among devices.

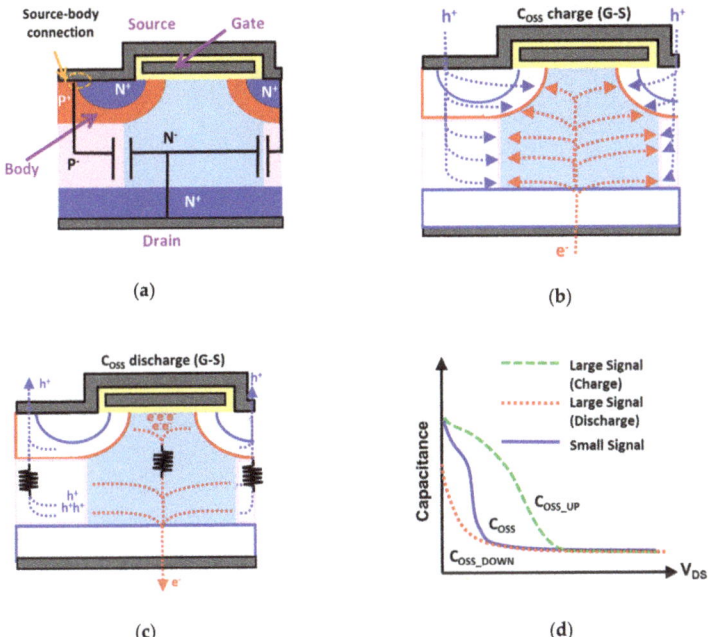

Figure 2. (a) Cross section of a Superjunction (SJ)-MOSFET basic cell. Description of (b) C_{OSS} charge and (c) C_{OSS} discharge. Electron (e-) and hole (h+) currents and charge pockets are indicated (red and blue). (d) Illustrative comparison between C_{OSS} extracted by small-signal (solid blue line) and large-signal (green dashed and red dotted lines).

E_i used to be neglected [14,15], but some simulations models started to consider non-linear C_{OSS} effects, and non-ZVS operation of SJ-MOSFETs [16–18]. However, C_{OSS} hysteresis discoveries are not still considered in those simulation models.

The degree of severity of C_{OSS} hysteresis varies from device to device depending on technological and geometrical features. Up to now, application notes and datasheets do not provide any information regarding this phenomenon. Besides, manufacturers only give small-signal characterization of the transistors, whereas C_{OSS} hysteresis is only detectable during large-signal analysis (Figure 2d). In order to solve this fact, a simple methodology to include in the datasheets enough information to generate simulation models predicting this behavior will be proposed.

In contrast to other reported techniques, a commercial system commonly used by power devices manufactures, an Auriga pulsed I–V system [19], is proposed. This characterization system is able to capture measurements with very high speed and resolution (up to 0.01% of max current), and it is temperature independent. Moreover, voltage/current measurements have emerged as the preferred method of capturing different characteristics of active devices. The simple setup and the voltage and current waveforms obtained using one of the SJ-MOSFETs under test are shown in Figure 3.

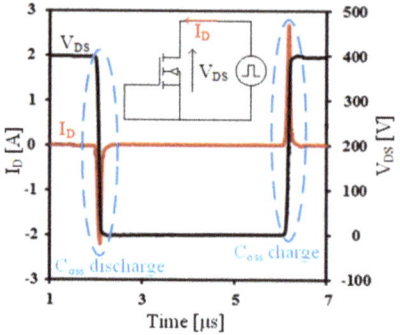

Figure 3. Auriga pulsed I–V tests: I_D and V_{DS} waveforms.

Using these voltage/current measurements, C_{OSS} large-signal curves during its charge and discharge can be inferred using

$$C_{OSS} = \frac{I_D}{\frac{dV_{DS}}{dt}}. \qquad (1)$$

Following the presented procedure, C_{OSS} large-signal curves during its charge and discharge of the SJ-MOSFETs under test (Table 1) were estimated (as an example, results of device under test 1 (DUT1) are included in Figure 4).

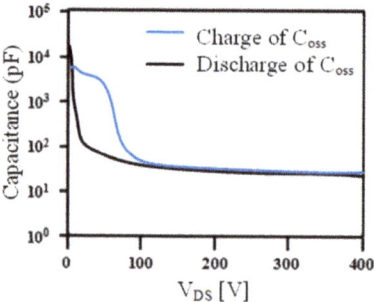

Figure 4. C_{OSS} large-signal curves of device under test 1 (DUT1) (Table 1) during its charge and discharge obtained using (1) and the Auriga pulsed I–V curves.

A detailed simulation model should include this behavior to obtain accurate simulation waveforms of the switching process. The calculated C_{OSS} could be described using a polynomial expression, but in this paper, the use of look-up-tables with pairs of values voltage-capacitance is proposed. Two different look-up-tables, one for the charge and one for the discharge, can be easily included in the spice model of the SJ-MOSFETs. This option is preferred (compared with polynomial expressions) from the point of view of saving computational time and the ease to use, follow, and change data if a different power device is chosen for simulation. The simulations results using the proposed model will be shown and compared with the experimental results in Section 4.

Table 1. List of Superjunction (SJ)-MOSFETs explicitly for LLC primary side with main electrical characteristics and figures of merit (FoMs). DUT, device under test.

DUT	Characteristics from Datasheet								FoM from Datasheet				Proposed FoM
	R_{ON} (mΩ)	BV_{DSS} (V)	R_G (Ω)	Q_G (nC)	Q_{GD} (nC)	Q_{GS} (nC)	E_{OSS} (μJ)[1]	Q_{OSS} (nC)[2]	$R_{ON} \cdot Q_G$ (Ω*nC)	$R_{ON} \cdot Q_{GD}$ (Ω*nC)	$R_{ON} \cdot E_{OSS}$ (Ω*μJ)	$R_{ON} \cdot Q_{OSS}$ (Ω*nC)	$R_{ON} \cdot E_i$ (Ω*μJ)
1	155	600	0.9	24	8	5	2.7	140	3.72	1.24	0.42	21.64	0.049
2	168	650	0.6	60	25	12	6.4	122	10.08	4.20	1.08	20.43	0.194
3	171	600	3.4	37	13	11	4.9	106	6.33	2.22	0.84	18.03	0.116
4	165	650	6	30	13	7.4	3.6	111	4.95	2.15	0.59	18.28	0.793
5	175	600	7	29	12	6	4.6	102	5.08	2.10	0.81	17.85	0.116
6	168	600	7	29	12	6	4.1	103	4.87	2.02	0.69	17.34	0.375

[1] C_{OSS} stored energy at V_{DS} = 400V. [2] Output characteristic charge, result of charging C_{OSS} (time-related effective output capacitance is considered) rising from 0 to 400 V.

3. Simple Methodology to Include E_i in the Datasheets

The cumulative energy (E_{CUM}) of C_{OSS} can be calculated with the previously shown voltage and current waveforms obtained using the Auriga pulsed I–V system.

$$E_{CUM} = \int I_D \cdot V_{DS}\ dt. \tag{2}$$

Using (2), the energy stored during the charge and extracted during the discharge of C_{OSS} can be easily estimated. In Figure 5, an example of the value of E_{CUM} using one of the SJ-MOSFETs under test is shown as an example. As can be seen, the stored energy is higher than the extracted energy, and this difference is the value of E_i. Concretely, E_i is considered as the energy losses after applying a complete cycle of charge–discharge to the device, and consequently is directly related to C_{OSS} hysteresis.

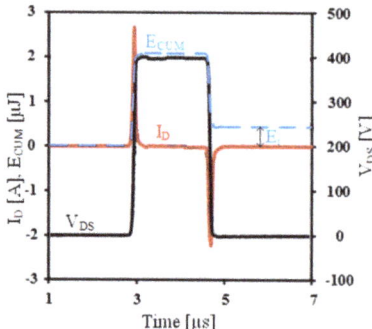

Figure 5. Auriga pulsed I–V tests: I_D and V_{DS} waveforms. Procedure to obtain the cumulative energy and the value of E_i to propose the new figure of merit (FoM).

The proposed FoM, defined as the conduction resistance (R_{ON}) multiplied by E_i, considers both R_{ON} (important for heavy loads) and E_i (crucial for low and medium loads), allowing a proper selection of SJ-MOSFETs in soft-switching applications operating at high frequencies. The lower the FoM value of an SJ-MOSFET, the higher its performance. In Section 4, the prediction of the performance of different SJ-MOSFETs using this FoM is validated with experimental efficiency results.

Besides, it is worth mentioning that, as occurring in other common FoMs, the direct and indirect proportionalities of R_{ON} and E_i with the die area result in an area-independent FoM. This is a preferred FoM approach to facilitate the benchmarking between technologies. In addition, common to other FoMs are the limitations for devices with a small die area, where the termination area could be as relevant as the active area of the transistor (R_{ON} does not perfectly scale with the die area).

4. Validation of the Proposed Simulation Model and FoM

The power supplies used in the applications mentioned in the introduction of this paper must comply with challenging standards, such as 80PLUS Titanium® [20]. An LLC resonant converter is the topology generally selected to develop this kind of power supply, mainly owing to their high efficiency and power density. More information and new models are needed to properly design these power converters operating at a very high switching frequency.

Silicon SJ-MOSFETs are the preferred devices during the design of the primary side of the LLC resonant converter as they meet the requirements regarding voltage, current, and frequency, and an accurate procedure for their proper selection for each specification is important, especially operating at a high switching frequency. The devices under test (DUT) in this work are detailed in Table 1. SJ-MOSFETs with similar voltage blocking capability, R_{ON}, and Q_{OSS} are selected in order to obtain a fair comparison under the same working conditions. In all the tests, ZVS is assured and, consequently, differences in the value of R_G are not relevant because the switching losses were forced to be independent of R_G. Exhaustive experimental tests are carried out using an LLC resonant converter with the DUTs. Waveforms, breakdown losses, and efficiency results are analyzed and compared.

4.1. LLC Resonant Converter Description

The previously described SJ-MOSFETs were tested in a commercial evaluation board of an LLC resonant converter [21] featuring the specifications of Table 2.

Table 2. LLC resonant converter evaluation board specifications.

Parameter	Value	Parameter	Value
Primary side devices	Si SJ-MOSFETs (DUTs)	Input Voltage, V_{IN} (V)	350–410
Secondary side devices	OptiMOS BSC010N04LS	Output voltage, V_{OUT} (V)	12
Gate driver IC	2EDL05N06PF	C_R (nF)	66
Maximum power (W)	600	L_R (uH)	15.5
Resonant frequency, f_{RES} (kHz)	157	L_M (uH)	195
Frequency range (kHz)	90–250	Transformer turns-ratio	16:1

The fundamental requirements related to a fixed resonant tank (C_R, L_R, and L_M) and deadtimes (t_D) are fulfilled, guaranteeing the ZVS inductive mode for the whole power range and for all the transistors under examination [13]. As the devices selected share very close values of R_{ON} and Q_{OSS}, there is no need to redesign different L_M values for each transistor, reassuring ZVS the whole load range. A simplified scheme of the LLC resonant converter with the main components and the sensing methods is shown in Figure 6.

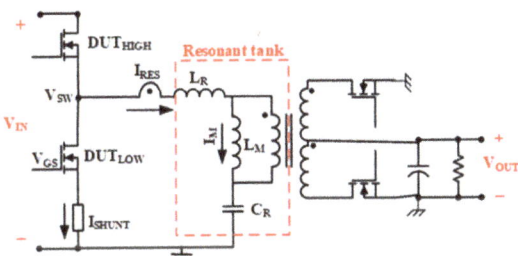

Figure 6. Simplified circuit scheme of the LLC resonant converter and sensing method.

Mixed-mode (MM) simulations were also carried out to analyze the evolution of certain signals that cannot be experimentally measured (as the current through the channel of the MOSFETs). The developed MM simulations consist of spice circuits, where the half-bridge (HB) structure of the primary side of the LLC resonant converter is replaced by TCAD (Technology Computer Aided Design) structures (finite-element structures) simulating the power transistors (DUT$_{HIGH}$ and DUT$_{LOW}$ in Figure 6).

Calibration of TCAD structures was done by means of process simulations in the case of own SJ-MOSFETs technologies, and by means of reverse engineering and reverse calibration technique in the case of other commercial SJ-MOSFETs technologies. Information about the doping profiles is included in the TCAD structures and data regarding voltages, power, magnetics, frequencies, and so on are extracted from the evaluation board datasheet [21].

The accuracy of the MM simulations and its good match with experimental waveforms can be seen in Figure 7. Moreover, the current through the channel of the DUT$_{LOW}$ (I_{CH}) obtained using MM simulation (it cannot be experimentally measured) was included to verify the ZVS operation (I_{CH} falls to zero before V_{SW} is increased). As I_{CH} is zero before V_{SW} rises, the area below P_{INS} waveform represents the energy stored in the output capacitance of the SJ-MOSFET during the turn-off (E_{off}). This energy cannot be considered as losses, because it can be retrieved in the turn-on.

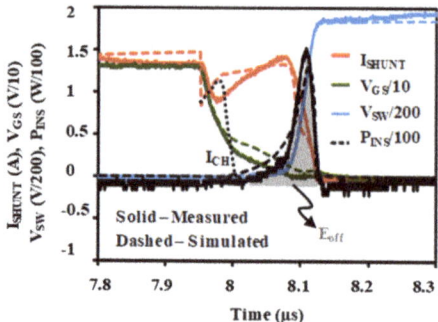

Figure 7. Simulated and measured waveforms for DUT_{LOW} during the turn-off. I_{SHUNT}, V_{GS}, and V_{SW} are referenced in Figure 6. P_{INS} is the instantaneous power (product of V_{SW} and I_{SHUNT}) and I_{CH} is the simulated current through the channel of the SJ-MOSFET.

4.2. Experimental Results, Efficiency, and Power Losses Break-Down

Several experimental measurements and waveforms are analyzed to validate the proper operation of the converter at different loads and with different DUTs. Examples of experimental waveforms measured in the converter are shown in Figure 8.

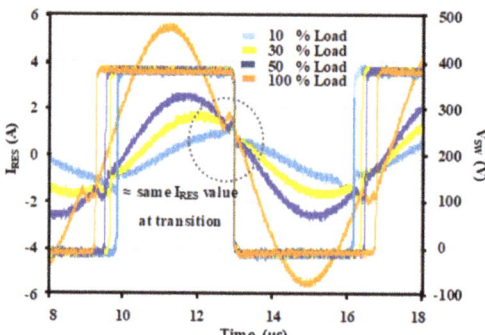

Figure 8. Experimental I_{RES} and V_{SW} measured at different loads for DUT1 as an example of how the resonant current varies with the load. As can be seen, different switching frequencies are also used for different loads.

In Figure 8, the current through the resonant tank (I_{RES} in Figure 6) is shown for different power levels, as well as its corresponding V_{SW} waveform. As can be seen, the I_{RES} value during the transition of both DUTs remains almost the same regardless of the load level, which will be helpful to estimate switching losses (they are calculated by means of the energy dissipated during the turn-on and turn-off) and to understand the behavior of the transistors during these transitions.

An efficiency comparison of the LLC resonant converter using all the DUTs as the primary side transistors is carried out in the whole power range, going from 10% to 100% of full load (600 W), always following the same test protocol and operating conditions. In order to minimize error measurements and its influence on the efficiency comparison, a repetitive protocol was performed using an automatic program based on Java. First, the converter is turned-on at 10% of maximum load, and it remains under this working condition for 15 minutes to achieve a constant working temperature. Then, the efficiency at 10% of maximum load is measured (this measurement is the result of averaging 10 consecutives measurements of V_{IN}, V_{OUT}, I_{IN}, and I_{OUT}). Finally, the load is increased, and new measurements are done after one minute. This procedure is repeated to 20%, 30%, 50%, 70%, and 100% of full load. In Figure 9, the differential efficiencies obtained are shown, taken as reference DUT1, as it is the device that shows best performance for the whole range.

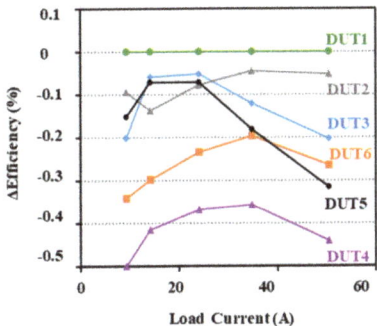

Figure 9. Differential efficiencies of the LLC resonant converter with respect to the best SJ-MOSFET (DUT1).

Using experimental waveforms of V_{GS}, V_{SW}, and I_{SHUNT}, switching (P_{SW}), driving (P_{DR}), and conduction (P_{ON}) power losses contributions from each DUT are calculated for three power loads demands of 60 W (10%), 300 W (50%), and 600 W (100%), and are shown in Figure 10a–c, respectively.

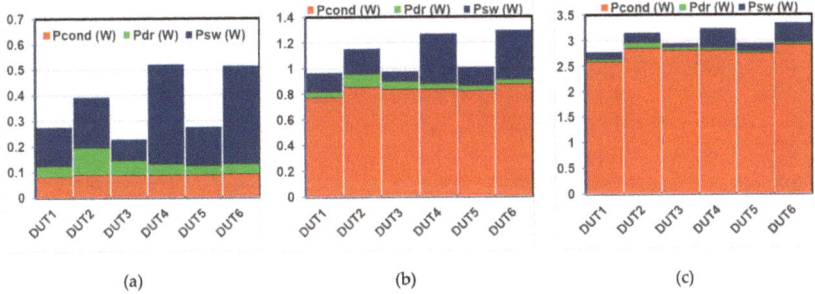

Figure 10. Measured power losses owing to driving (P_{DR}), switching (P_{SW}), and conduction (P_{ON}) at (a) 10%, (b) 50%, and (c) 100% load for each primary-side SJ-MOSFET.

In Figure 10a, at a low load, whereas low P_{ON} losses remain almost equal for all the DUTs, differences in P_{DR} losses have a small impact and P_{SW} losses are dominant. For heavy loads (Figure 10c), P_{ON} is by far the main factor of losses in the SJ-MOSFETs, yet disparity among the P_{SW} losses is discernible. At a medium load (Figure 10b), divergence in P_{SW} among transistors makes the difference (P_{ON} losses are the highest, but fairly the same value, but differences at P_{SW} have a great impact in the losses contribution). Even performing ZVS, P_{SW} losses are relevant and differences in the total power losses between DUTs are the result of $P_{SW} + P_{DR}$ at light loads (Figure 10a) and $P_{SW} + P_{ON}$ at heavy loads (Figure 10c). These P_{SW} losses under ZVS conditions are consistent with the existence of the E_i previously reported.

4.3. Validation of the Simulation Model Including C_{OSS} Hysteresis

The developed spice model of all the SJ-MOSFETs under test includes the definition of the C_{OSS} with two look-up-tables with pairs of values voltage-capacitance, one for the charge and one for the discharge (obtained using the procedure presented in Section 2). On the basis of the circuit proposed in Figure 6 and using the proposed spice models of the SJ-MOSFETs, some simulations of the LLC resonant converter are carried out using LTSpice. In these simulations, emphasis is on the primary side of the converter and the accurate definition of the C_{OSS} value. Experimental and simulated VSW waveforms are compared in Figure 11 and good agreement is obtained.

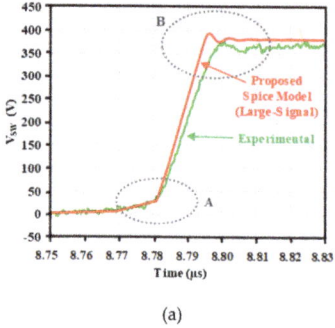
(a)

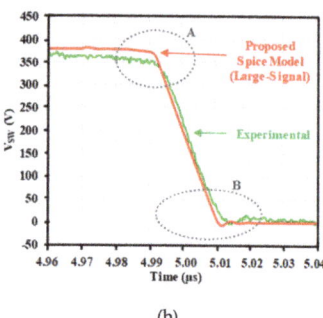
(b)

Figure 11. V_{SW} waveform extracted by experimental measurement (green) and simulation with the proposed large-signal spice model (red) during (**a**) the increase of V_{SW} and (**b**) the decrease of V_{SW}.

It should be noted that the equivalent capacitance seen from the port defined by V_{SW} is the parallel combination of the output capacitance of DUT_{LOW} and DUT_{HIGH} (two nonlinear capacitors), defined as

$$C_{eq} = \frac{C_{OSS_{DUT_{LOW}}}(V_{SW}) \cdot C_{OSS_{DUT_{HIGH}}}(V_{IN} - V_{SW})}{C_{OSS_{DUT_{LOW}}}(V_{SW}) + C_{OSS_{DUT_{HIGH}}}(V_{IN} - V_{SW})} \tag{3}$$

Taking into account that the value of C_{OSS} of each SJ-MOSFET is different during its charge and discharge, C_{eq} is not symmetric (as presented in previous works not including the C_{OSS} hysteresis [15]) and a different value is obtained when V_{SW} goes up and down. As can be seen in Figure 12, the equivalent impedance during the increase of V_{SW} (C_{eq1}) has the same value at high voltage than the equivalence impedance during the decrease of V_{SW} (C_{eq2}) at low voltage. Consequently, in Figure 11, similar V_{SW} evolution can be seen in the corners marked as A and B during the increase and the decrease of V_{SW}. The proposed spice model captures the corner asymmetry (see corners A and B have different curvature) when V_{SW} goes up and down during DUT_{LOW} turn-off and turn-on transitions, thus being consistent with the existence of a C_{OSS} hysteresis and matching the experimental measurements.

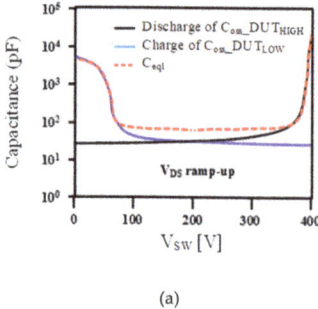
(a)

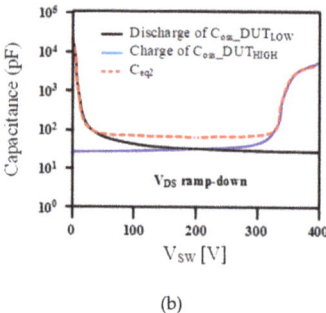
(b)

Figure 12. Derivation of C_{eq1} and C_{eq2}, which are asymmetric with respect to the charge and discharge of the C_{OSS} of DUT_{HIGH} and DUT_{LOW}. (**a**) Increase of V_{SW} and (**b**) decrease of V_{SW}.

4.4. Validation of the Proposed FoM including E_i

In Figure 9, there is not a clear trend regarding the efficiency that DUTs show for different load demands. Some of them might be suitable for low power, but, in contrast, their performance is worse at full load. That is the case of DUT5. FoMs based on the information provided by the datasheet do not always explain these differences in operation. For example, the worse performance at full load of DUT5 can be explained by its high on-resistance, but the performance for a light load of DUT5 is better than the performance of DUT6, while their switching characteristics are almost the same (even a bit better than those of DUT6).

Consequently, new FoMs are needed to know in detail where the power losses come from, as a great percentage of the converter total losses is attributable to the primary-side SJ-MOSFETs [13] for the whole load range. The proposed FoM should allow a proper selection of the SJ-MOSFETs in an LLC resonant converter. In the last column of Table 1, the value of the proposed FoM for all the DUTs is included, while in Figure 13, these values are compared with respect to DUT1.

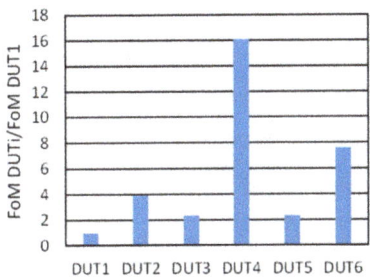

Figure 13. Comparison of the proposed FoM of the DUTs.

The best performance of the LLC is obtained with DUT1, which also has the lowest value of the proposed FoM. DUT2, DUT3, and DUT5 have low values of the proposed FoM and the performance of the LLC with them is also good, especially at medium and light loads. DUT4 has the highest value of the proposed FoM, and the LLC with this DUT has the lowest performance.

As can be clearly seen, the proposed FoM can predict the better performance of the LLC with DUT5 than with DUT6 (especially at light load), which cannot be explained using the characteristics from datasheets.

5. Conclusions

The existence of E_i in power devices, which produces significant switching energy losses in high switching frequency power converters, even under ZVS, has been shown in this paper. Moreover, E_i are related to a C_{OSS} hysteresis.

A simple methodology using a commercial system is suggested to manufacturers to provide E_i (which is not included in datasheets) and information about the C_{OSS} hysteresis. The relevance of the C_{OSS} hysteresis information is validated by developing simulation models that accurately match the experimental charge and discharge waveforms of the C_{OSS}. These new models will allow the designer to better predict the behavior of the power devices and their corresponding power losses, in order to to be able to design more efficient applications.

Efficiency experimental results on an LLC resonant converter are used to validate the suitability of the proposed FoM including E_i, to properly select the transistors used in soft-switching power converters operating at high frequencies.

The impact of these kinds of losses is important in high switching frequency power converters and should be properly modelled to be able to predict the performance of different commercial power transistors in case the designer needs to compare several of those for a certain application; consequently, new data and models are needed.

Author Contributions: Conceptualization, J.R.; methodology, M.R.R.; software, G.G.; validation, M.R.R.; formal analysis, A.R.; resources, P.V.; writing–original draft preparation, M.R.R. and A.R.; writing–review and editing, A.R., D.G.L., and J.R.; supervision, A.R., J.R., and D.G.L.; All authors have read and agreed to the published version of the manuscript.

Funding: This research was supported by the Spanish Government under projects MCIU-19-RTI2018-099682-A-I00 and by the Principado de Asturias under the project FC-GRUPIN-IDI/2018/000179

Conflicts of Interest: The authors declare no conflict of interest.

Nomenclature

V_{IN}	input voltage of the LLC converter.
V_{OUT}:	output voltage of the LLC converter
DUT_{HIGH}/DUT_{LOW}	primary side MOSFETs under test
L_R	discrete series resonant inductance of the resonant tank
L_M	magnetizing inductance of the resonant tank
C_R	discrete resonant capacitor of the resonant tank
V_{GS}	gate-to-source voltage of the DUT_{LOW}
V_{SW}	drain-to-source voltage of the DUT_{LOW}. Switching voltage
I_{RES}	current through the resonant tank
I_{SHUNT}	drain-to-source current through DUT_{LOW}
I_{CH}	simulated current through the channel of the DUT_{LOW}
P_{INS}	instantaneous power dissipated in the DUT_{LOW}. Product of I_{SHUNT} and V_{SW}.
C_{OSS}	non-linear output capacitance of the MOSFETs
G, S, and D	gate, source and drain terminals of the MOSFETs
P	p-doped zone
N	n-doped zone
E_i	intrinsic energy losses after charging and discharging C_{OSS} up to a certain drain-to-source voltage (V_{DS}) in off-state
E_{CUM}	cumulative energy in the MOSFET when applying a voltage pulse on it during off-state.
$P_{SW}, P_{DR},$ and P_{ON}	switching, driving, and conduction power losses of the MOSFETs

References

1. Chen, Y.; Wang, H.; Liu, Y.F. Improved hybrid rectifier for 1-MHz LLC-based universal AC-DC adapter. In Proceedings of the IEEE Applied Power Electronics Conference and Exposition (APEC), Tampa, FL, USA, 26–30 March 2017; pp. 23–30.
2. Deng, J.; Li, S.; Hu, S.; Mi, C.C.; Ma, R. Design methodology of LLC resonant converters for electric vehicle battery chargers. *IEEE Trans. Veh. Technol.* **2014**, *63*, 1581–1592. [CrossRef]
3. Hu, H.; Fang, X.; Chen, F.; Shen, Z.J.; Batarseh, I. A modified high-efficiency LLC converter with two transformers for wide input-voltage range applications. *IEEE Trans. Power Electron.* **2013**, *28*, 1946–1960. [CrossRef]
4. Fedison, J.B.; Fornage, M.; Harrison, M.J.; Zimmanck, D.R. COSS related energy loss in power MOSFETs used in zero-voltage-switched applications. In Proceedings of the IEEE Applied Power Electronics Conference and Exposition (APEC), Fort Worth, TX, USA, 16–20 March 2014; pp. 150–156.
5. Li, X.; Bhalla, A. Comparison of intrinsic energy losses in unipolar power switches. In Proceedings of the IEEE Wide Bandgap Power Devices and Applications (WiPDA), Fayetteville, AR, USA, 7–9 November 2016; pp. 228–232.
6. Tong, Z.; Zulauf, G.; Rivas-Davila, J. A Study on off-state losses in silicon-carbide schottky diodes. In Proceedings of the IEEE 19th Workshop on Control and Modeling for Power Electronics (COMPEL), Padua, Italy, 25–28 June 2018.

7. Rothmund, D.; Bortis, D.; Kolar, J.W. Accurate Transient Calorimetric Measurement of Soft-Switching Losses of 10-kV SiC mosfets and Diodes. *IEEE Trans. Power Electron.* **2018**, *33*, 5240–5250. [CrossRef]
8. Zulauf, G.; Tong, Z.; Rivas-Davila, J. Considerations for active power device selection in high- and very-high-frequency power converters. In Proceedings of the IEEE 19th Workshop on Control and Modeling for Power Electronics (COMPEL), Padua, Italy, 25–28 June 2018.
9. Fedison, J.B.; Harrison, M.J. Coss hysteresis in advanced superjunction mosfets. In Proceedings of the IEEE Applied Power Electronics Conference and Exposition (APEC), Long Beach, CA, USA, 20–24 March 2016; pp. 247–252.
10. Roig, J.; Bauwens, F. Origin of anomalous coss hysteresis in resonant converters with superjunction FETs. *IEEE Trans. Electron Devices* **2015**, *62*, 3092–3094. [CrossRef]
11. Zulauf, G.D.; Roig-Guitart, J.; Plummer, J.D.; Rivas-Davila, J.M. Coss measurements for superjunction MOSFETs: Limitations and opportunities. *IEEE Trans. Electron Devices* **2019**, *66*, 578–584. [CrossRef]
12. Roig, J.; Gomez, G.; Bauwens, F.; Vlachakis, B.; Rogina, M.R.; Rodriguez, A.; Lamar, D.G. High-accuracy modelling of ZVS energy loss in advanced power transistors. In Proceedings of the IEEE Applied Power Electronics Conference and Exposition (APEC), San Antonio, TX, USA, 4–8 March 2018; pp. 263–269.
13. Maria, R.; Rogina, A.; Rodriguez, D.; Lamar, G.; Roig, J.; Vanmeerbeek, P.; Bauwens, F. Novel selection criteria of primary side transistors for LLC resonant converters. In Proceedings of the IEEE Control and Modelling for Power Electronic (COMPEL), Padova, Italy, 25–28 June 2018.
14. Elferich, R. General ZVS half bridge model regarding nonlinear capacitances and application to LLC design. In Proceedings of the IEEE Energy Conversion Congress and Exposition (ECCE), Edinburgh, UK, 15–20 September 2012; pp. 4404–4410.
15. Costinett, D.; Maksimovic, D.; Zane, R. Circuit-oriented treatment of nonlinear capacitances in switched-mode power supplies. *IEEE Trans. Power Electron.* **2015**, *30*, 985–995. [CrossRef]
16. Christen, D.; Biela, J. Analytical switching loss modeling based on datasheet parameters for mosfets in a half-bridge. *IEEE Trans. Power Electron.* **2019**, *34*, 3700–3710. [CrossRef]
17. Miftakhutdinov, R. New ZVS analysis of PWM converters applied to super-junction, GaN and SiC power FETs. In Proceedings of the IEEE Applied Power Electronics Conference and Exposition (APEC), Charlotte, NC, USA, 15–19 March 2015; pp. 336–341.
18. Kasper, M.; Burkart, R.M.; Deboy, G.; Kolar, J.W. ZVS of power MOSFETs revisited. *IEEE Trans. Power Electron.* **2016**, *31*, 8063–8067. [CrossRef]
19. Available online: https://focus-microwaves.com/pulsed-iv/ (accessed on 13 April 2019).
20. Available online: http://www.80PLUS.org (accessed on 13 April 2019).
21. Steiner, A.; di Domenico, F.; Catly, J.; Stückler, F. *600 W Half Bridge LLC Evaluation Board with 600 V CoolMOS™ C7*; Infineon Technology: Neubiberg, Germany, 2015.

© 2020 by the authors. Licensee MDPI, Basel, Switzerland. This article is an open access article distributed under the terms and conditions of the Creative Commons Attribution (CC BY) license (http://creativecommons.org/licenses/by/4.0/).

MDPI
St. Alban-Anlage 66
4052 Basel
Switzerland
Tel. +41 61 683 77 34
Fax +41 61 302 89 18
www.mdpi.com

Energies Editorial Office
E-mail: energies@mdpi.com
www.mdpi.com/journal/energies

www.ingramcontent.com/pod-product-compliance
Lightning Source LLC
LaVergne TN
LVHW070703100526
838202LV00013B/1020